DBKDA 2025

The Seventeenth International Conference on Advances in Databases, Knowledge, and Data Applications

March 9th –13th, 2025

Lisbon, Portugal

DBKDA 2025 Editors

Constantine Kotropoulos, Aristotle University of Thessaloniki, Greece

Peter Kieseberg, St. Pölten University of Applied Sciences, Austria

Printed from e-media with permission by:

Curran Associates, Inc.
57 Morehouse Lane
Red Hook, NY 12571

Some format issues inherent in the e-media version may also appear in this print version.

International Academy, Research, and Industry Association (IARIA)
412 Derby Way
Wilmington, DE 19810

Phone: (408) 893-6407
Fax: (408) 527-6351

petre@iaria.org

Additional copies of this publication are available from:

Curran Associates, Inc.
57 Morehouse Lane
Red Hook, NY 12571 USA
Phone: 845-758-0400
Fax: 845-758-2633
Email: curran@proceedings.com
Web: www.proceedings.com

Table of Contents

DBKDA 2025

Foreword

The Seventeenth International Conference on Advances in Databases, Knowledge, and Data Applications (DBKDA 2025), held between March 9 - 13, 2025, continued a series of international events covering a large spectrum of topics related to advances in fundamentals on databases, evolution of relation between databases and other domains, data base technologies and content processing, as well as specifics in applications domains databases.

Advances in different technologies and domains related to databases triggered substantial improvements for content processing, information indexing, and data, process and knowledge mining. The push came from Web services, artificial intelligence, and agent technologies, as well as from the generalization of the XML adoption.

High-speed communications and computations, large storage capacities, and load-balancing for distributed databases access allow new approaches for content processing with incomplete patterns, advanced ranking algorithms and advanced indexing methods.

Evolution on e-business, ehealth and telemedicine, bioinformatics, finance and marketing, geographical positioning systems put pressure on database communities to push the 'de facto' methods to support new requirements in terms of scalability, privacy, performance, indexing, and heterogeneity of both content and technology.

We take here the opportunity to warmly thank all the members of the DBKDA 2025 Technical Program Committee, as well as the numerous reviewers. The creation of such a high quality conference program would not have been possible without their involvement. We also kindly thank all the authors who dedicated much of their time and efforts to contribute to DBKDA 2025. We truly believe that, thanks to all these efforts, the final conference program consisted of top quality contributions.

Also, this event could not have been a reality without the support of many individuals, organizations, and sponsors. We are grateful to the members of the DBKDA 2025 organizing committee for their help in handling the logistics and for their work to make this professional meeting a success.

We hope that DBKDA 2025 was a successful international forum for the exchange of ideas and results between academia and industry and for the promotion of progress in the fields of databases, knowledge and data applications.

We are convinced that the participants found the event useful and communications very open. We also hope that Nice provided a pleasant environment during the conference and everyone saved some time for exploring this beautiful city

DBKDA 2025 Chairs:

DBKDA 2025 Steering Committee

Fritz Laux, Reutlingen University, Germany
Andreas Schmidt, Karlsruhe Institute of Technology / University of Applied Sciences
Erik Hoel, Esri, USA
Lisa Ehrlinger, Software Competence Center Hagenberg GmbH, Austria
Peter Kieseberg, St. Pölten University of Applied Sciences, Austria
Constantine Kotropoulos, Aristotle University of Thessaloniki, Greece

DBKDA 2025 Publicity Chairs

Francisco Javier Díaz Blasco, Universitat Politècnica de València, Spain
Ali Ahmad, Universitat Politècnica de València, Spain
Sandra Viciano Tudela, Universitat Politecnica de Valencia, Spain
José Miguel Jiménez, Universitat Politecnica de Valencia, Spain

DBKDA 2025

Committee

DBKDA 2025 Steering Committee

Fritz Laux, Reutlingen University, Germany
Andreas Schmidt, Karlsruhe Institute of Technology / University of Applied Sciences
Erik Hoel, Esri, USA
Lisa Ehrlinger, Software Competence Center Hagenberg GmbH, Austria
Peter Kieseberg, St. Pölten University of Applied Sciences, Austria
Constantine Kotropoulos, Aristotle University of Thessaloniki, Greece

DBKDA 2025 Publicity Chairs

Francisco Javier Díaz Blasco, Universitat Politècnica de València, Spain
Ali Ahmad, Universitat Politècnica de València, Spain
Sandra Viciano Tudela, Universitat Politecnica de Valencia, Spain
José Miguel Jiménez, Universitat Politecnica de Valencia, Spain

DBKDA 2025 Technical Program Committee

Taher Omran Ahmed, University of Technology and Applied Sciences, Ibri, Oman / Alzintan University, Libya
Julien Aligon, Institut de Recherche en Informatique de Toulouse (IRIT) | Université Toulouse 1 Capitole, France
Alaa Alomoush, University Malaysia Pahang, Malaysia
Emmanuel Andres, Hôpitaux Universitaires de Strasbourg, France
Vincenzo Arceri, Università degli Studi di Parma, Italy
Zeyar Aung, Masdar Institute of Science and Technology, UAE
Qiushi Bai, Microsoft, USA
Aruna Bansal, IBM India Pvt. Ltd., India
Nelly Barret, Politecnico di Milano, Italy
Christian Beecks, University of Hagen, Germany
Giacomo Bergami, Newcastle University, UK
Jam Jahanzeb Khan Behan, Université libre de Bruxelles (ULB), Belgium / Universidad Politécnica de Cataluña (UPC), Spain
Flavio Bertini, University of Parma, Italy
Vincenzo Bonnici, University of Parma, Italy
Savong Bou, University of Tsukuba, Japan
Ali Boukehila, University of Annaba, Algeria
Zouhaier Brahmia, University of Sfax, Tunisia
Martine Cadot, LORIA, Nancy, France
Alessandro Castelnovo, Intesa Sanpaolo S.P.A / University of Milano Bicocca, Italy
Basabi Chakraborty, Iwate Prefectural University, Japan / Madanapalle Institute of Technology and Science, India
Sanjay Chaudhary, Ahmedabad University, India
Yung Chang Chi, National Cheng Kung University, Taiwan

Richard Chbeir, Université de Pau et des Pays de l'Adour (UPPA), France
Jong Choi, Oak Ridge National Laboratory, USA
Stefano Cirillo, University of Salerno, Italy
Alessia Auriemma Citarella, University of Salerno, Italy
Miguel Couceiro, LORIA, France
Malcolm Crowe, University of the West of Scotland, UK
Fabiola De Marco, University of Salerno, Italy
Monica De Martino, Istituto per la Matematica Applicata e Tecnologie Informatiche "Enrico Magenes" | Consiglio Nazionale delle Ricerche, Italy
Bipin C. Desai, Concordia University, Montreal, Canada
Luigi Di Biasi, University of Salerno, Italy
Marianna Di Gregorio, University of Salerno, Italy
Ivanna Dronyuk, Jan Dlugosz University in Czestochowa, Poland
Cedric du Mouza, CNAM (Conservatoire National des Arts et Métiers), Paris, France
Lisa Ehrlinger, Senior Researcher at the Information Systems Research Group, Germany
Amir Hajjam El Hassani, University of Technology of Belfort Montbeliard, France
Gledson Elias, Federal University of Paraíba (UFPB), Brazil
Austen Fan, University of Wisconsin-Madison, USA
Matteo Francia, University of Bologna, Italy
Iwao Fujino, Tokai University, Japan
Satvik Garg, University of Rochester, USA
Ana González-Marcos, Universidad de La Rioja, Spain
Gregor Grambow, Hochschule Aalen, Germany
Luca Grilli, University of Foggia, Italy
Binbin Gu, University of California, Irvine, USA
Boujemaa Guermazi, Toronto Metropolitan University, Canada
Robert Gwadera, Cardiff University, UK
Mohammed Hamdi, Najran University, Saudi Arabia
Tobias Hecking, German Aerospace Center (DLR), Germany
Mohammad Rezwanul Huq, East West University, Bangladesh
Hamidah Ibrahim, Universiti Putra Malaysia, Malaysia
Vladimir Ivančević, University of Novi Sad, Serbia
Ivan Izonin, Lviv Polytechnic National University, Ukraine
Marouen Kachroudi, Université de Tunis El Manar, Tunisia
Aida Kamisalic Latific, University of Maribor, Slovenia
Saeed Kargar, University of California, Santa Cruz, USA
Jeyhun Karimov, Huawei Munich Research Center, Germany
Tahar Kechadi, University College Dublin (UCD), Ireland
Maqbool Khan, Pak-Austria Fachhochschule - Institute of Applied Sciences and Technology, Haripur, Pakistan
Mourad Khayati, University of Fribourg, Switzerland
Daniel Kimmig, solute GmbH, Germany
Sotirios I. Kontogiannis, University of Ioannina, Greece
Constantine Kotropoulos, Aristotle University of Thessaloniki, Greece
Katrien Laenen, KU Leuven University, Belgium
Prarit Lamba, Intuit, USA
Jean-Charles Lamirel, Université de Strasbourg | LORIA, France
Nadira Lammari, CEDRIC-Cnam, France

Copyright Information

Exploring Latent Concepts in SHAP Values

- A New Approach Using Singular Value Decomposition -

Yukari Shirota
Faculty of Economics
Gakushuin University
Tokyo, Japan
e-mail: yukari.shirota@gakushuin.ac.jp

Tmaki Sakura
Nikkei Economics Centre

Tokyo, Japan
e-mail: sakura@jcer.or.jp

Abstract— **In this paper, we introduce a novel explainable AI method called "SHAP_SVD" for regression analysis. The Shapley value, originally developed by Lloyd Shapley, has gained prominence as a key tool in explainable AI (XAI) through its adaptation as SHAP by Lundberg. In regression analysis, SHAP values are computed using characteristic functions of the data, representing the contribution of each explanatory variable to the target value. Our proposed SHAP_SVD method applies Singular Value Decomposition (SVD), a dimensionality reduction technique, to the SHAP value matrix. The eigenvalues and eigenvectors extracted via SVD capture the core structure of the SHAP matrix, revealing "concepts" or "latent semantic concepts." In SVD, these concepts are represented by two sets of eigenvectors. As a case study, we demonstrate the regression analysis of stock price growth rates for Indian and Japanese automakers, where two principal concepts were identified, consistently reflected across both sets of eigenvectors.**

Keywords- *XAI; Shapley values; SHAP; Singular Value Decomposition; India automakers.*

I. INTRODUCTION

EXplainable AI (XAI) has emerged as a critical field, bridging the gap between complex machine learning models and human interpretability. Among the numerous XAI techniques developed, Shapley values, introduced by Lloyd Shapley, have gained prominence for their ability to allocate the contribution of each feature in a model's predictions [1-3]. Adapted into the SHapley Additive exPlanations (SHAP) framework by Lundberg [4-6], this method has become a widely used tool for interpreting machine learning models, particularly in regression analysis [7-10]In regression analysis, SHAP values quantify the contribution of each explanatory variable to the target value by utilizing characteristic functions of the data. These values offer deep insights into feature importance and interaction. However, as the complexity and dimensionality of the data increase, the interpretation of SHAP values becomes challenging. Traditional SHAP methods are limited in their ability to reveal underlying structures within the data, especially when dealing with high-dimensional or multi-faceted variables.

To address this limitation, we propose a novel method, SHAP_SVD, which applies Singular Value Decomposition (SVD) to the SHAP value matrix. SVD, a well-known dimensionality reduction technique, captures the core structure of a matrix by decomposing it into eigenvalues and eigenvectors [11-13]. This allows us to extract latent semantic concepts or "principal components" from the SHAP matrix. By leveraging SVD, SHAP_SVD uncovers these underlying concepts, represented by two sets of eigenvectors, thus providing a richer understanding of the relationships between explanatory variables and the target variable.

As a concrete example, we apply SHAP_SVD to the regression analysis of stock price growth rates for Indian and Japanese automakers, using the market capitalization growth rates as the target variable. Through this analysis, we identified two key latent concepts, which we refer to as (1) Balanced (Well-balanced) type, and (2) Sales Growth Rate (SGR)-driven type," extracted from the SHAP_SVD decomposition. By plotting company data on a two-dimensional plane defined by these two principal component axes, we are able to conduct a detailed analysis of the characteristics driving market capitalization growth for each company. This approach enables us to visualize and understand the underlying factors that influence the stock price performance of companies in both markets.

The remainder of this paper is organized as follows. In Section 2, we describe the data used for the analysis, including the data sources. Section 3 explains the methods applied, introducing both the SHAP analysis and our proposed SHAP_SVD method. Section 4 presents the SHAP results, analyzing the contributions of explanatory variables to the target values. Section 5 details the SHAP_SVD method, illustrating how Singular Value Decomposition is applied to the SHAP matrix and how latent concepts are extracted. In Section 6, we discuss existing work related to explainable AI and dimensionality reduction, comparing these approaches with our proposed method. Section 7 provides a discussion of the results and their implications. Finally, Section 8 concludes the paper with a summary of contributions and suggestions for future research.

II. DATA

In this section, we shall explain the regression data. In the regression, we use Market Capitalization (MC) data. MC amount is a stock price times the number of issued stocks. The target variable is the Indian and Japanese automakers' "**annual MC growth rates**" in 2022. The MC growth rate in year XXX is defined as ***(MC_XXX – MC_(XXX-1)) / (MC_(XXX-1))***, namely, the ratio based on the previous year. We would like to find the dominant factors for the rapid MC growth rates. The MC data we used were retrieved from the ORBIS company database by Bureau van Dijk, the last data update date being 2024/06/22.

The damages caused by COVID-19 have revealed vulnerable supply chains in automakers. This regression frame assumes that the competence of supply chains and new market development are prerequisites for the long-run sustenance of companies' high business performance, leading to high stock price evaluation [14]. Therefore, we select four managerial factors as the explanatory variables. Sales Growth Ratio (SGR) represents the new market development competence, and FArate represents the supply chain competence [14-16]. The tangible Fixed Asset amount (FA) is the third explanatory variable used to identify the impact of the firm's scalability. These factors allow companies to earn satisfactory levels of profitability, such as their stock prices, Return On Equity (ROE), and Return On Assets (ROA). In addition, we focus on labor productivity. Labor productivity in the manufacturing sector refers to the goods or value one worker produces within a specific period. It is a crucial metric for assessing the efficiency and competitiveness of a manufacturing operation. We want to evaluate which is more significant on the target tangible assets or labor productivity. Labor productivity was calculated here using the following formula:

$$Labor\ Productivity = \frac{Total\ Value\ Added}{Number\ of\ Workers} = \frac{Net\ Sales - Cost\ of\ Goods\ Sold}{Number\ of\ Workers}$$

The managerial index data of the automobile companies were also retrieved from the ORBIS company database. After removing companies with missing annual data, the number reached 67, including 11 Indian and 56 Japanese automakers. We conducted the regressions with the data.

III. METHODS

In this section, the methods we used are described. The flow chart of the analysis is as follows:

1. **XGBoost Regression**: The given data is input into the XGBoost Regressor [17], and then the regression function f(X) is generated as output.
2. **SHAP Evaluation**: Based on the regression function f(X), SHAP values for each data are calculated. In this case study, we use four explanatory variables and 67 companies, resulting in a SHAP matrix of size 67 x 4.
3. **SVD of SHAP Matrix**: Applying SVD to the SHAP matrix M, the decomposition outputs three matrices such that $M=U\Sigma V^T$.
4. **SHAP_SVD Interpretation**: The eigenvectors and eigenvalues extracted from SVD are interpreted to uncover underlying concepts. The two sets of eigenvectors are referred to as CompanyEigenVectors and SHAP_Eigenvectors, representing different two viewpoints of the underlying concepts.

SHAP is a method based on Shapley values from cooperative game theory, designed to explain machine learning model predictions, including those in regression tasks. A key strength of SHAP is its ability to create a characteristic function for the data, allowing it to calculate the contribution of each feature based on the characteristics of individual data points. This ensures that the contribution of each feature to the model's output is computed fairly and additively. SHAP enhances the interpretability of complex models, offering insights into how specific data characteristics influence predictions. We used XGBoost as the regression algorithm.

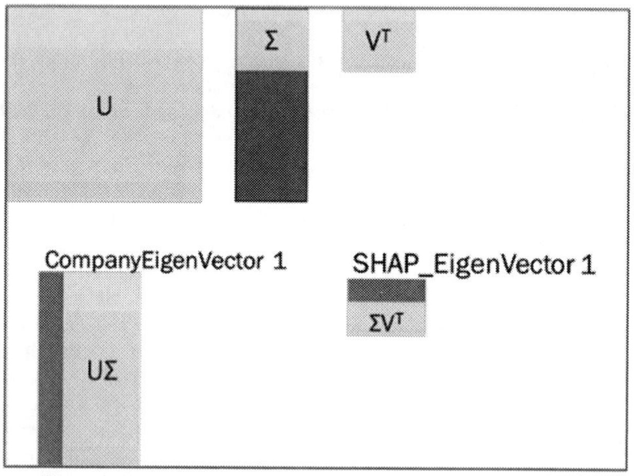

Figure 1. CompanyEigenVectors and SHAP_EigenVectors in SVD.

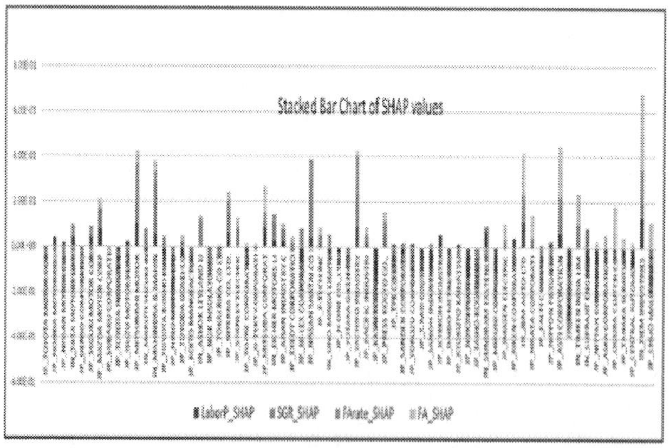

Figure 2. Stacked bar chart of the SHAP values.

In step 3, the SHAP matrix M is decomposed by SVD. As shown in Figure 1, the SHAP matrix M is decomposed to the matrixes, U, Σ, and V. The shape of U is squared.

The matrix Σ is padded by 0 in the blue area in Figure 1. The shape of U times Σ becomes a rectangle. The first row of this rectangle becomes CompanyEigenVector_1. CompanyEigenVector_1 corresponds to the first EigenVector. The shape of Σ times V^T becomes a rectangle. The first line of this rectangle becomes SHAP_EigenVector_1. SHAP_EigenVector _1 corresponds to the first EigenVector.

In step 4, SHAP_SVD interpretation, the two kinds of eigenvectors are evaluated, which are named in this case study CompanyEigenVector and SHAP_EigenVector (see the red parts in Figure 1).

IV. SHAP RESULTS

In this section, the results of the SHAP evaluation are presented.

The regression model, developed using XGBOOST, achieved an R-squared value exceeding 0.99, indicating a highly accurate fit to the data. Using the regression model f(X), the characteristic function is approximately evaluated. SHAP values are found based on the characteristic function. Figure 2 shows the SHAP values. The horizontal line shows the company IDs. Figure 2 illustrates a stack bar chart of SHAP values of the individual companies. There in each company has four SHAP values corresponding to the four explanatory variables. The horizontal zero line shows the average target value of the companies. The sum of four SHAP values in each company becomes its deviation of the target value from the average. The SHAP matrix M size becomes 67 times 4 which is the target matrix of the SVD.

V. SHAP_SVD METHOD

In this section, SHAP_SVD method is explained.

After the SVD of the SHAP matrix, the singular value (SV) lists are obtained (see Figure 3). The SVs express the strength of the latent concepts. The ratio is approximately 9:6:4:3.

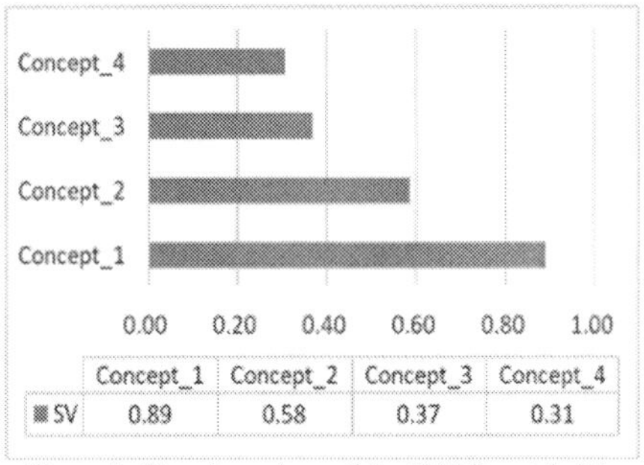

Figure 3. Singular values of the SHAP matrix M.

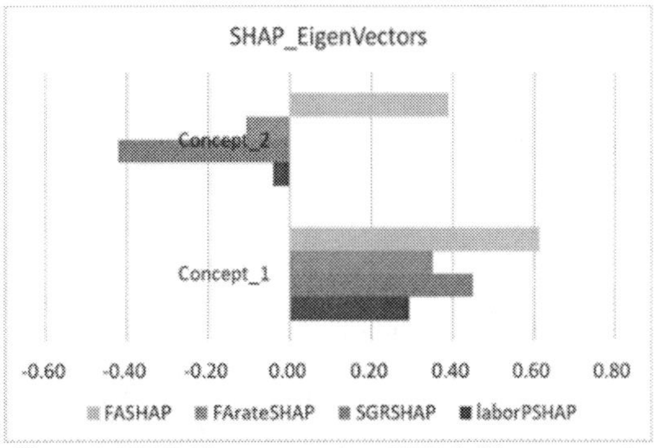

Figure 4. SHAP_Eigenvectors_1 and SHAP_Eigenvectors_2.

Then, we will interpret the meaning of the concepts, focusing on the two largest SVs. The concepts can be interpreted by the two aspects. First, using SHAP_EigenVectors, the concepts will be expressed in Figure 4. The bottom bar graph presents the SHAP_EigenVector_1 with four elements. Our interpretation of the two concepts is as follows:

- 1st concept: All elements cooperate and have high values, with particularly high SHAP for FA.
- 2nd concept: SGR_SHAP is high, and FA's SHAP is low (expressing it this way reverses the sign of the vector elements).

We name the concepts (1) Balanced (Well-balanced) type and (2) Sales Growth Rate (SGR)-driven type.

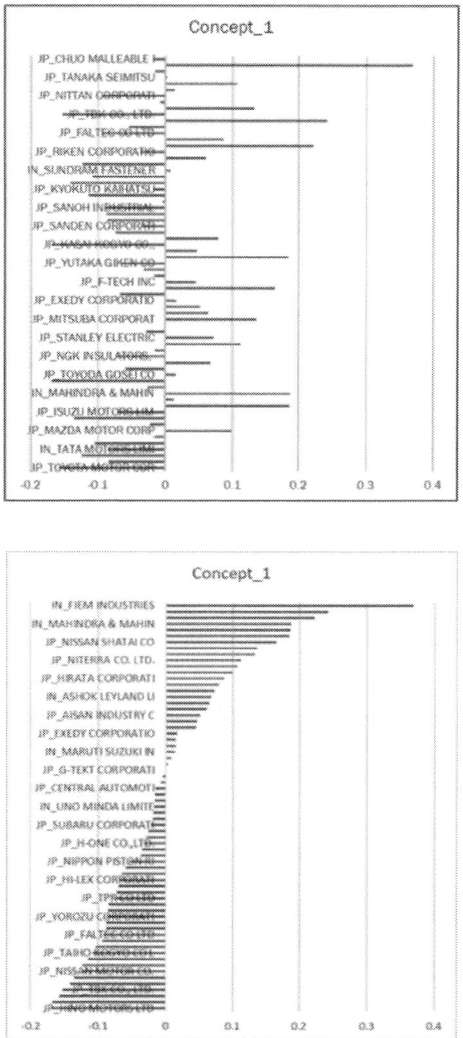

Figure 5. CompanyEigenVector_1 (the bottom is the sorted one).

Then, using CompanyEigenVectors, the concepts will be interpreted. Figure 5 shows CompanyEigenVector_1. The bottom graph is the sorted version. The largest element company was FIEM Industries. FIEM is a well-established company in India, primarily known for its expertise in automotive lighting. With over 50 years of experience, FIEM has grown into a leading supplier for Original Equipment Manufacturers (OEMs) in India and abroad. In the representation using CompanyEigenVectors, the first concept can be interpreted as companies with SHAP distributions similar to FIEM.

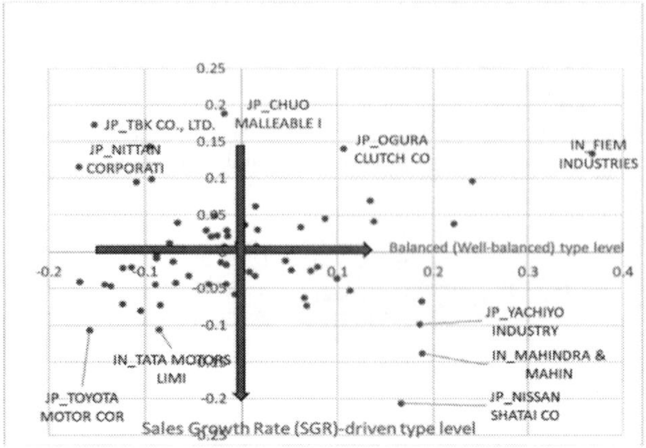

Figure 6. A scattering plot of CompanyEigenVector_1 elements and CompanyEigenVector_2 element values of 67 companies

Figure 6 shows a scattering plot of CompanyEigenVector_1 element values and CompanyEigenVector_2 element values of 67 companies. The representatives concerning CompanyEigenVector_2 elements, which measure the SGR-driven level, will be interpreted. Figure 7 shows the first SHAP values of the SGR-driven type level's highest five companies. As shown in Figure 6, the first second principle component (y-axis) is oriented downwards, and the company with the highest y-value is Nissan Shatai, followed by Mahindra as the second highest.

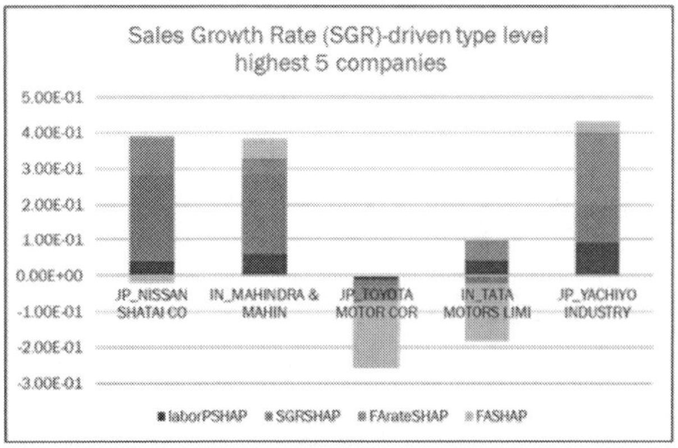

Figure 7. SHAP values of the SDG-driven type level highest five companies.

In the three companies like Nissan Shatai and Mahindra, where SHAP values are positive, the SGR_SHAP (Sales Growth Rate SHAP) is large while FA_SHAP (Fixed Assets SHAP) is small. This suggests sales growth is the main driving factor rather than the size of tangible fixed assets. For two companies with negative SHAP values, the impact of SGR_SHAP dragging performance is smaller (almost zero), compared to the negative impact of FA_SHAP.

Toyota, for example, typically the strength is tangible fixed assets, with FA_SHAP being large and positive when the target value is positive and negative when the target is negative.

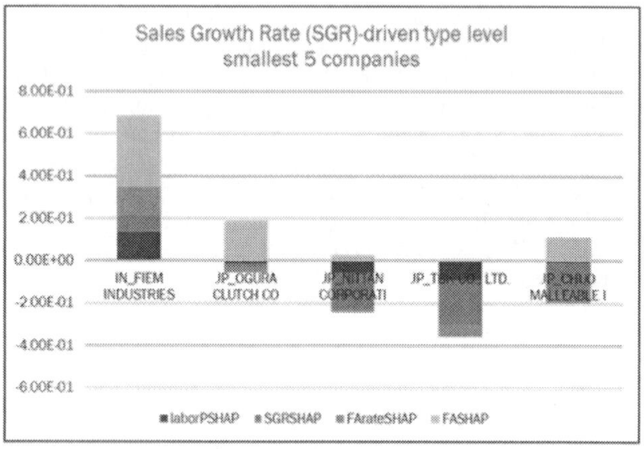

Figure 8. SHAP values of the SDG-driven type level smallest five companies.

Next, we will evaluate SHAP values of the SDG-driven type level smallest five companies (see Figure 8). In the three companies with positive SHAP values, FA_SHAP (Fixed Assets SHAP) is large and SGR_SHAP (Sales Growth Rate SHAP) is small, indicating that the companies are driven more by the size of their tangible assets rather than sales growth. For the two companies with negative SHAP values, SGR_SHAP is dragging performance less than FA_SHAP, with FA_SHAP values being close to zero. This suggests that FA has minimal impact in these cases.

VI. Existing Work

In this section, the related existing works are presented. The first allocation field is a stock price evaluation and the second allocation field is text mining.

Random Matrix Theory (RMT) and portfolio

RMT has been applied to stock market analysis to reduce noise in financial data. RMT helps distinguish real market signals from random fluctuations in stock price correlations [18]-[23]. The flow charts of the method are as follows:

1. Correlation Matrix: Begin by calculating the correlation matrix of stock returns. This matrix sizes the number of companies times the number of sales dates.
2. RMT Filtering: RMT is used to separate meaningful signals from random noise. Eigenvalues of the correlation matrix are compared with theoretical RMT predictions. Larger eigenvalues represent true market information, while smaller ones reflect noise.
3. Singular Value Decomposition (SVD): SVD is applied to further clean the correlation matrix, focusing on the

significant components. This improves the matrix's accuracy, filtering out noise.

4. Portfolio Optimization: Using the noise-reduced correlation matrix, more accurate risk and return estimates can be made, improving portfolio construction.

Latent Semantic Analysis (LSA)

LSA is a widely used technique in Natural Language Processing (NLP), primarily for analyzing semantic relationships between documents. It is often applied in tasks such as topic modeling, semantic analysis, and information retrieval [19]-[25].

Overview of LSA:

1. Purpose: LSA aims to convert the semantic relationships between words and documents into a lower-dimensional latent semantic space, allowing for the identification of similarities and relationships between documents. This helps uncover hidden patterns or topics within the text.
2. Method: LSA begins by creating a co-occurrence matrix that captures how often words appear together in a document. This matrix models the relationships between words and documents. Then, SVD is applied to reduce the dimensionality of the matrix. By using SVD, LSA compresses the high-dimensional data while preserving the important semantic relationships and filtering out noise.

LSA is a powerful mathematical approach for interpreting the semantic structure of text and is utilized in search engines, automatic summarization systems, document clustering, and more. SVD techniques are mathematically explained in [11, 13]. The two kinds of EigenVectors and the relationship among the three decomposed matrixes are clearly explained using visualization in [19, 20].

VII. Discussion

In this section, we will discuss the result. The objective of the analysis is a grouping of companies. The proposed SHAP_SVD method can extract the essence of the given SHAP value matrix. In the paper, the two extracted concepts were (1) Balanced (Well-balanced) type, and (2) Sales Growth Rate (SGR)-driven type. Using each CompanyEigenVectors' element values, we can measure each company's (1) Balanced (Well-balanced) type level and (2) Sales Growth Rate (SGR)-driven type level. As shown in Figure 6, the scattering plot of the companies by CompanyEigenVectors can uncover the individual companies' characteristics. The horizontal axis represents the Balanced (Well-balanced) type level. These higher-level companies can be divided by the vertical axis into two groups, which are an "SGR_SHAP higher and FA_SHAP lower" group and a "FA_SHAP higher and SGR_SHAP lower" group. This means that these companies exhibit a similar pattern of feature contributions, reflecting a particular type of balance or focus in their business models.

SHAP values can reflect each company's characteristics more accurately than using the raw input data. Therefore, analyzing SHAP values through SVD (Singular Value Decomposition) allows for more accurate dimensionality reduction based on the characteristics of each company. This method enhances the ability to capture distinct business drivers by compressing the data in a way that aligns with each company's unique attributes, offering deeper insights compared to standard SHAP analysis. In corporate management, creating appropriate Key Performance Indexes (KPIs) is crucial. The EigenVectors (principal component axes) derived from SHAP_SVD analysis can serve as the first step in developing these KPIs. By identifying the most important factors influencing business performance through dimensionality reduction, SHAP_SVD helps to highlight key metrics that align with a company's unique characteristics, providing a strong foundation for effective KPI creation.

VIII. Conclusions

In this paper, we proposed the SHAP_SVD method, which combines SHAP values with SVD for regression analysis. The SHAP_SVD method enables the extraction of core concepts from the SHAP value matrix, providing a more accurate representation of each company's characteristics compared to using raw input data. In our case study, we identified two main concepts: (1) Balanced (Well-balanced) type and (2) SGR-driven type. By analyzing the SHAP values through SVD, we were able to group companies based on their unique feature contributions, revealing distinct business drivers.

This method offers deeper insights into corporate data by aligning the dimensionality reduction process with the specific characteristics of each company. Furthermore, the eigenvectors derived from SHAP_SVD can serve as a foundation for developing effective KPIs, helping businesses to identify and focus on the factors that most significantly influence their performance.

Acknowledgment

The research was partly supported by the fund of Gakushuin University Computer Centre project 2024.

References

[1] A. E. Roth, "Introduction to the Shapley value," *The Shapley value*, pp. 1-27, 1988.

[2] A. E. Roth, *The Shapley value: essays in honor of Lloyd S. Shapley*. Cambridge University Press, 1988.

[3] L. S. Shapley, "A value for n-person games, Contributions to the Theory of Games, 2, 307–317," ed: Princeton University Press, Princeton, NJ, USA, 1953.

[4] S. M. Lundberg, G. G. Erion, and S.-I. Lee, "Consistent individualized feature attribution for tree ensembles," *arXiv preprint arXiv:1802.03888*, 2018.

[5] S. M. Lundberg and S.-I. Lee, "A unified approach to interpreting model predictions," *Advances in neural information processing systems*, vol. 30, 2017.

[6] S. M. Lundberg and S.-I. Lee, "Consistent feature attribution for tree ensembles," *arXiv preprint arXiv:1706.06060*, 2017.

[7] A. R. Javed, W. Ahmed, S. Pandya, P. K. R. Maddikunta, M. Alazab, and T. R. Gadekallu, "A survey of explainable artificial intelligence for smart cities," *Electronics*, vol. 12, no. 4, p. 1020, 2023.

[8] R. Dwivedi *et al.*, "Explainable AI (XAI): Core ideas, techniques, and solutions," *ACM Computing Surveys*, vol. 55, no. 9, pp. 1-33, 2023.

[9] A. Chaddad, J. Peng, J. Xu, and A. Bouridane, "Survey of explainable AI techniques in healthcare," *Sensors,* vol. 23, no. 2, p. 634, 2023.

[10] Y. Shirota, K. Kuno, and H. Yoshiura, "Time series analysis of shap values by automobile manufacturers recovery rates," in *Proceedings of the 2022 6th International Conference on Deep Learning Technologies*, 2022, pp. 135-141.

[11] C. M. Bishop and N. M. Nasrabadi, *Pattern recognition and machine learning* (no. 4). Springer, 2006.

[12] T. Hastie, R. Tibshirani, J. H. Friedman, and J. H. Friedman, *The elements of statistical learning: data mining, inference, and prediction*. Springer, 2009.

[13] S. Theodoridis and K. Koutroumbas, *Pattern recognition*. Elsevier, 2006.

[14] Y. Shirota, M. Fujimaki, E. Tsujiura, M. Morita, and J. A. D. Machuca, "A SHAP Value-Based Approach to Stock Price Evaluation of Manufacturing Companies," in *2021 4th International Conference on Artificial Intelligence for Industries (AI4I)*, 2021: IEEE, pp. 75-78.

[15] M. Fujimaki, E. Tsujiura, and Y. Shirota, "Automobile Manufacturers Stock Price Recovery Analysis at COVID-19 Outbreak," in *6th World Conference on Production and Operations Management – P&OM Nara 2022*, Nara, Japan, 2022: EurOMA (European Operations Management Association), p. Decision Science Institute Best Paper Award.

[16] K. Yamaguchi, "Relationship Analysis Between Stock Prices and Financial Statements in the Automobile Industry," in *2023 14th IIAI International Congress on Advanced Applied Informatics (IIAI-AAI)*, 2023: IEEE, pp. 442-445.

[17] XGBoostDevelopers. "XGBoost Documentation (Revision 534c940a.)." https://xgboost.readthedocs.io/en/stable/ (accessed 2022/11/13.

[18] V. Plerou, P. Gopikrishnan, B. Rosenow, L. A. N. Amaral, T. Guhr, and H. E. Stanley, "Random matrix approach to cross correlations in financial data," *Physical Review E,* vol. 65, no. 6, p. 066126, 2002.

[19] Y. Shirota and B. Chakraborty, "Visual Explanation of Eigenvalues and Math Process in Latent Semantic Analysis," *Information Engineering Express, Information Engineering Express,* vol. 2, no. 1, pp. 87-96, 2016. [Online]. Available: http://www.iaiai.org/journals/index.php/IEE/article/view/70.

[20] Y. Shirota and B. Chakraborty, "Visual explanation of mathematics in Latent semantic analysis," in *2015 IIAI 4th International Congress on Advanced Applied Informatics*, 2015: IEEE, pp. 423-428.

Route Planning in Wildfire Areas by Integrating a Modified A* Algorithm with Deep Learning

Manavjit Singh Dhindsa
Department of Electrical and Computer Engineering
University of Waterloo, Waterloo, Canada
e-mail: ms2dhind@uwaterloo.ca

Kshirasagar Naik, Pin-Han Ho
Department of Electrical and Computer Engineering
University of Waterloo, Waterloo, Canada
e-mail: {snaik | p4ho}@uwaterloo.ca

Marzia Zaman
Research and Development
Cistel Technology, Ottawa, Canada
e-mail: marzia@cistel.com

Chung-Horng Lung
Department of Systems and Computer Engineering
Carleton University, Ottawa, Canada
e-mail: chlung@sce.carleton.ca

Srinivas Sampalli
Department of Computer Scienceg
Dalhousie University, Halifax, Canada
e-mail: srini@cs.dal.ca

Abstract—**Wildfires pose a significant threat to life, property, and ecosystems, with their frequency and intensity escalating due to climate change. Effective evacuation planning is critical to mitigating wildfire impacts, yet it remains a challenging task in dynamic, high-risk scenarios. This paper presents a framework for safe path planning that integrates wildfire spread predictions from state-of-the-art deep learning models with an optimized A* (OA*) algorithm. The proposed approach utilizes binary fire masks to generate safe and efficient evacuation routes while adhering to strict safety constraints, such as maintaining buffer zones around fire-affected regions. Experimental results show the algorithm's capability to generate actionable paths and accurately identify no-path scenarios under diverse wildfire conditions. This framework offers a robust solution for real-time evacuation planning, contributing to the broader efforts of wildfire management and disaster mitigation.**

Keywords-Route Navigation; Deep Learning; Forest fire; A Algorithm; Path Planning; Wildfire Prediction; Machine Learning.*

I. Introduction

Wildfires, once a natural mechanism for maintaining ecological balance, have transformed into a global environmental and socio-economic crisis. Historically, fires served essential ecological roles, such as clearing dead vegetation and recycling nutrients. However, in recent decades, climate change, combined with evolving fire management policies, has led to a marked increase in both the frequency and intensity of wildfires. Projections suggest that the annual occurrence of very large fires (greater than 5000 hectares) could quadruple between 2041 and 2070 compared to the period from 1971 to 2000 [1]. For example, the 2020 wildfire season in the United States burned over 4 million hectares, including the devastating August Complex fire, which consumed more than 400,000 hectares [2]. These fires caused approximately $19 billion in economic losses and emitted 112 million metric tons of carbon, surpassing the annual emissions of all vehicles in California combined [3].

The repercussions of wildfires extend far beyond the immediate destruction of ecosystems. They threaten biodiversity, water resources, carbon storage, and air quality, while also imposing significant burdens on public health and the economy. Infrastructure losses, increased insurance costs, and healthcare challenges due to smoke exposure are just a few examples of their cascading effects [4], [5].

Given these challenges, there is an urgent need for effective wildfire management strategies, particularly in emergency response scenarios. One crucial component of this is the ability to generate safe evacuation routes based on accurate wildfire spread predictions. Predictive models provide critical insights into the spatial and temporal dynamics of wildfire behavior, enabling the identification of high-risk areas. However, translating these predictions into actionable evacuation plans requires robust path-planning algorithms that can navigate dynamic and hazardous environments.

This paper focuses on the development of a safe path planning framework that utilizes wildfire spread predictions to identify secure evacuation routes. Using results from deep learning models, such as U-Net and Vision Transformers (ViT), which generate binary fire masks representing fire-affected zones, this framework leverages a modified A* pathfinding algorithm to compute safe paths in real time. The proposed approach integrates the outputs of these prediction models into a graph-based search process, ensuring that routes avoid hazardous areas while adhering to safety constraints.

By addressing the challenge of converting wildfire predictions into actionable safe paths, this work contributes to the broader goal of improving emergency response capabilities. The framework demonstrates the potential to support timely and efficient evacuation planning, ensuring the safety of individuals and minimizing the impacts of wildfires on affected communities.

The remainder of this paper is organized as follows: Section II reviews related work on wildfire prediction and pathfinding techniques. Section III describes the methodology, focusing on the use of prediction results from Deep Learning models integrated with a modified A* algorithm for safe pathfinding. Section IV presents the experimental setup, results, and evaluation. Section V concludes with key findings and future research directions.

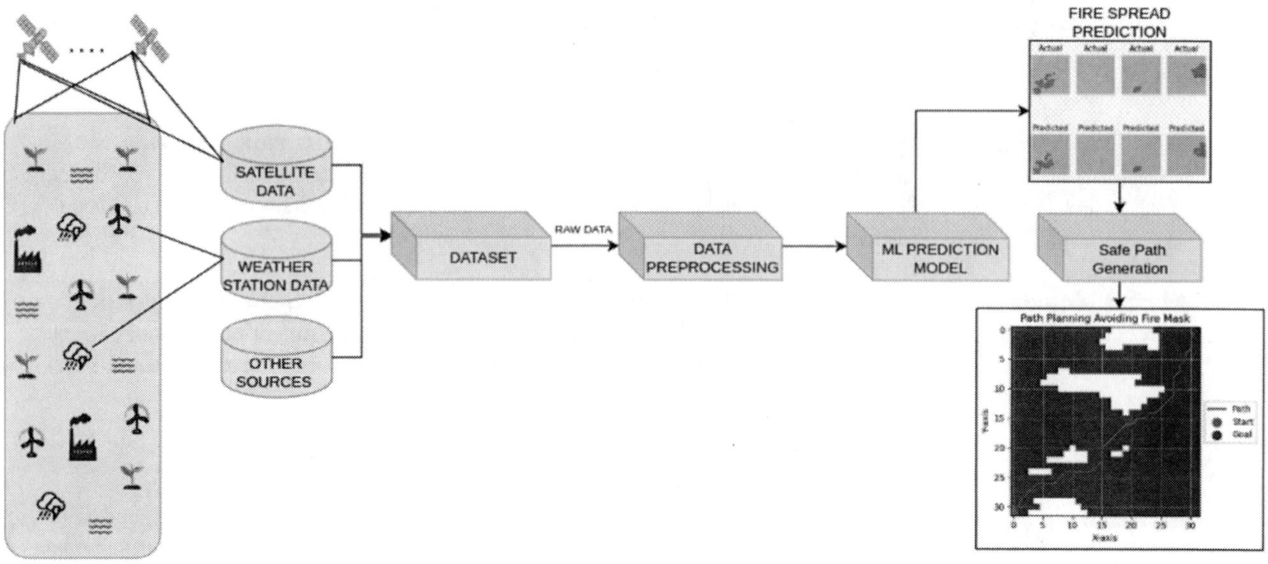

Figure 1: System Model

II. Related Work

Traditionally, wildfire prediction models have been based on empirical, physics-based, and cellular automata methods. Empirical models, like the Rothermel fire spread model, used historical fire data to predict fire intensity and rate of spread [6]. Physics-based models, such as BEHAVE [7] and FARSITE [8], simulate fire behavior by incorporating physical processes, like wind, temperature, and fuel type. Cellular automata models further enhanced fire dynamics simulation by introducing stochastic and spatially explicit models, such as those by Karafyllidis and Thanailakis [9]. While these traditional models were effective, they struggled with the complexity of fire spread dynamics, particularly in non-linear environments. The advent of machine learning (ML) and deep learning (DL) introduced new capabilities. ML models, including Random Forest and Support Vector Machines [10], have demonstrated success in capturing the intricate relationships between various factors influencing fire spread, whereas DL models like CNNs [11] and U-Nets [12] excel in processing satellite imagery and capturing both spatial and temporal patterns. These advancements offer significantly improved predictions by modeling the complex dynamics of wildfire propagation with higher accuracy and robustness than earlier methods.

For deep learning models to effectively predict wildfire spread, robust and multimodal datasets are essential. Datasets like WildfireDB [13] and WildfireSpreadTS [14] combine data from multiple sources, providing essential training foundations despite resolution and data quality challenges. These datasets enable DL models to learn from vast amounts of data, improving the predictive power by accounting for the multifaceted nature of fire behavior. Their ability to integrate spatial, temporal, and environmental data is fundamental in overcoming the limitations of earlier models, enabling better forecasting and resource allocation in wildfire management.

Path planning in dynamic environments, particularly during wildfires, is critical for ensuring the safety of evacuees and first responders. Dijkstra's algorithm, developed in 1959, laid the foundation for finding optimal paths in static environments [15]. However, for dynamic and uncertain environments like wildfires, A* algorithm is more suitable due to its integration of heuristics, which optimize pathfinding while considering obstacles and risk factors [16]. The A* algorithm provides a balance between computational efficiency and path optimality, making it a popular choice for evacuation planning. Moreover, its ability to consider dynamic factors, such as moving hazards or fire progression, allows for real-time adjustments, ensuring timely and safe routes.

For safe evacuation during wildfires, modified versions of path planning algorithms, such as those proposed by Wang et al. [17] and Xu et al. [18], integrate dynamic hazard modeling, risk assessment, and real-time fire prediction data. These modified algorithms adapt to the constantly changing conditions of wildfire environments, ensuring that the generated paths remain safe despite the unpredictability of fire spread. Systems that use real-time fire prediction data, can significantly improve the accuracy of evacuation plans. By continuously updating the predicted spread of the fire, these systems enable the generation of optimal evacuation routes for both the general public and emergency responders, minimizing risks during fire emergencies.

III. Methodology

Wildfires present an unpredictable and highly dynamic threat to life, property, and ecosystems. With accurate predictions of wildfire spread, it becomes possible to plan and execute evacuation strategies. In this work, we focus on the

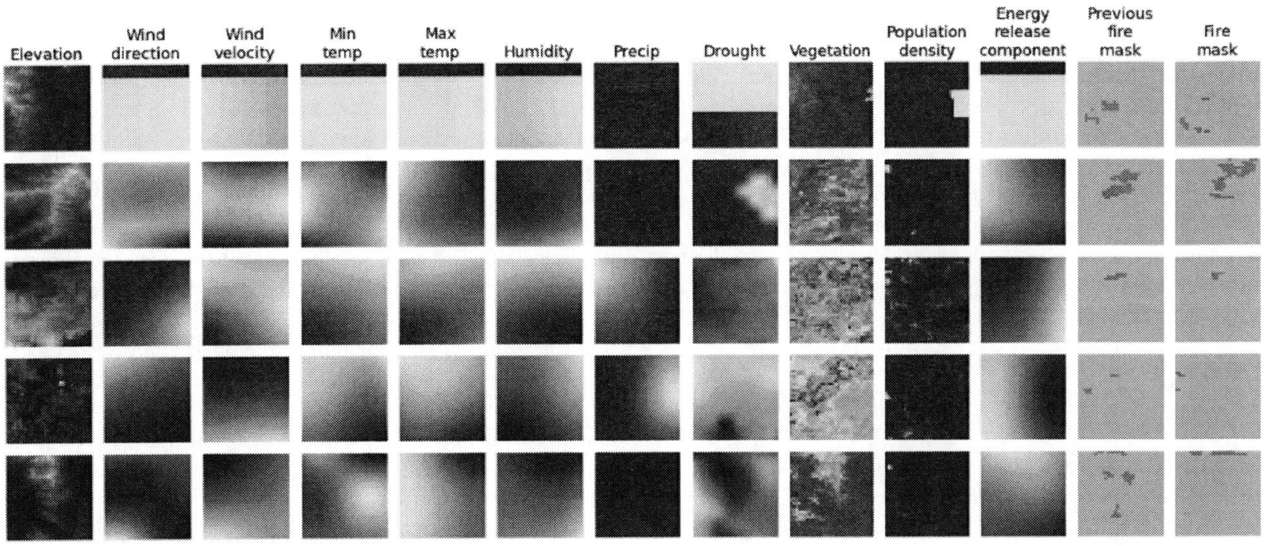

Figure 2: Example images from the dataset showing each of 13 features derived from various sources.

challenge of developing a robust safe path planning framework that integrates results from wildfire spread predictions to generate safe routes for evacuation.

A. *Wildfire Spread Prediction*

The process of generating a 2D wildfire spread prediction begins by integrating diverse data sources, such as vegetation indices, meteorological data (e.g., temperature, humidity, and wind speed), topographical features, and past fire occurrences. The multimodal inputs are preprocessed to ensure consistency, forming a cohesive dataset for model training (Fig. 2).

The dataset is processed by a deep learning (DL) model in two stages: first, extracting spatial and temporal features to capture relationships like environmental influences on fire propagation; second, identifying complex interactions that traditional methods struggle to model.

The DL model outputs a 2D fire mask prediction for the next day, representing the likelihood of wildfire spread. This binary or probabilistic map indicates fire-affected areas at a grid-cell level. The predicted fire mask can be post-processed for visualization or used directly in downstream applications, such as path planning and resource allocation. This workflow highlights how DL models effectively integrate diverse data to produce actionable insights.

B. *Problem Formulation*

The prediction of wildfire spread has been achieved using deep learning models such as U-Net and Vision Transformers (ViT). Our task is to leverage these predictions to develop evacuation paths that avoid areas affected by fire. The wildfire predictions are provided as binary grids, where each grid cell indicates whether a particular area is affected by fire or not. The primary challenge is to navigate safely through fire-prone areas using wildfire predictions while ensuring safety and efficiency. This challenge is addressed by formulating the problem as a graph-based search on binary grids

The safe path planning problem involves finding a path from a designated start point (such as the current location of individuals) to a safe destination while avoiding fire-affected areas. Mathematically, we represent the problem as a graph-based search problem, where the grid of wildfire predictions serves as a map. Each grid cell is treated as a node, and the goal is to find a safe route from the start node to the destination node, navigating through non-burning areas.

Let the grid be represented by the binary map $\mathbf{Y} \in \mathbb{R}^{H \times W}$, where H and W are the height and width of the grid, and each element $y_{(h,w)}$ is a binary value: $y_{(h,w)} = 1$ for fire-affected areas and $y_{(h,w)} = 0$ for non-burning areas. The objective is to use a modified A algorithm to find the optimal path, which avoids the cells marked as 1 (fire) and minimizes the distance while maximizing safety.

We define the safe path planning problem as follows:

$$\mathbf{P}_{\text{safe}} = \min_{\mathbf{P}} \left(\sum_{i=1}^{n} d_{i,i+1} \right); \quad y_{(h,w)} = 0, \forall (h,w) \in \mathbf{P} \quad (1)$$

where $\mathbf{P}$ is the path from the start to the destination, $d_{i,i+1}$ is the distance between consecutive nodes along the path, and the condition $y_{(h,w)} = 0$ ensures that all nodes along the path avoid fire-affected areas.

By integrating the results of the wildfire spread predictions into this path-planning process, we aim to provide an automated, safe, and efficient evacuation strategy for areas under threat. The modified A* algorithm, optimized for this context, will use both the fire predictions and traditional route planning criteria (e.g., distance, time, accessibility) to generate the safest and most viable evacuation paths.

The fire predictions are used as input for the optimized A*(OA*) algorithm to implement safe evacuation routes. The fire predictions, represented as binary masks, guide the A* algorithm in finding safe evacuation routes and avoiding fire-affected zones. Fig. 3 illustrates how a complex spiral-shaped fire mask is converted into a binary grid, which acts as a platform for applying the OA* algorithm and generating a corresponding fire mask with pixel values representing "fire" or "no fire." This approach integrates deep learning predictions with pathfinding algorithms to provide actionable, real-time evacuation strategies.

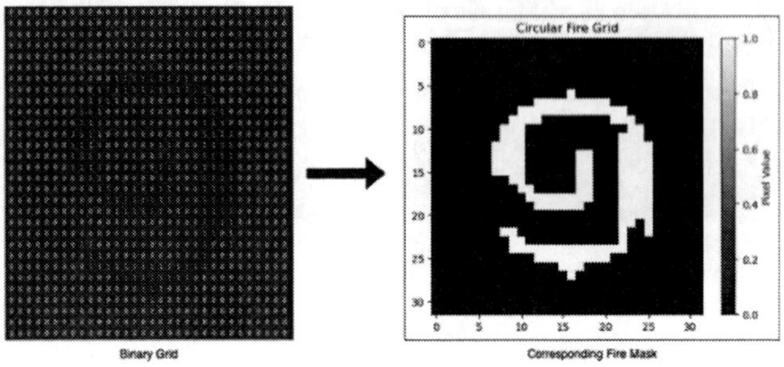

Figure 3: Binary grid converted into corresponding Fire Mask for application of Safe Path Planning.

C. *Proposed Algorithm: Optimized A* for Safe Path Planning*

The proposed safe path planning algorithm enhances the traditional A* search algorithm to dynamically navigate through wildfire-affected regions, ensuring that the generated paths avoid hazardous fire zones. This modified A* algorithm uses the output of the wildfire spread prediction model, represented as binary fire masks, as input. These fire masks encode areas affected by fire as '1' (danger zones) and safe zones as '0'. The objective is to generate the safest, shortest possible path from a start node to a goal node while avoiding fire regions and minimizing risk.

At a high level, the Optimized A* (OA*) algorithm integrates essential components to enable safe path planning. The core components include a heuristic function, a safety evaluation function, and a path reconstruction function. The heuristic function computes the Euclidean distance between nodes, estimating the cost of the shortest path to the goal:

$$f(n) = g(n) + h(n) \tag{2}$$

where $g(n)$ represents the cost from the start node to the current node, and $h(n)$ is the heuristic estimate of the cost to the goal. This function guides the algorithm by prioritizing nodes with the lowest estimated total cost.

The *IsSafe* function evaluates the safety of each node by inspecting its surrounding area within a predefined buffer. For each candidate node, this function iterates over neighboring cells to check whether they fall within fire-affected zones or exceed the grid boundaries. Nodes deemed unsafe are excluded from the search, ensuring the path avoids direct or proximate exposure to fire hazards.

The *ReconstructPath* function traces the final route from the goal node back to the start node. It utilizes a map of predecessors, which records the most efficient path during the search, and iteratively builds the route in reverse, ensuring the shortest and safest path is returned upon successful completion of the algorithm.

The OA* algorithm operates iteratively, beginning with the initialization of the priority queue, which tracks nodes based on their $f(n)$ values. The algorithm starts with the input fire mask, the start and goal nodes, and the specified safety buffer. For each node, the algorithm considers all eight possible movement directions, including diagonals, and evaluates the safety and cost of each neighboring node. If a neighboring node is safe and offers a lower cost than previously recorded, it is added to the open set for further exploration. Unsafe nodes, as determined by the *IsSafe* function, are pruned from the search space.

When a neighboring node satisfies the safety and cost criteria, its $g(n)$ value is updated to reflect the traversal cost, and its $f(n)$ value is recomputed. If the goal node is reached, the algorithm terminates, and the *ReconstructPath* function maps out the shortest, safest route. If the priority queue is exhausted without finding a path, the algorithm concludes that no viable route exists, ensuring that no unsafe recommendations are made.

This integration of heuristic evaluation, safety constraints, and efficient path reconstruction ensures that the OA* algorithm generates paths that are both optimal in terms of distance and safe for traversal. The approach is particularly effective in scenarios requiring real-time decision-making, such as emergency evacuations, firefighting operations, and autonomous navigation in fire-affected regions. By prioritizing safety while maintaining efficiency, the OA* algorithm addresses the critical challenges posed by dynamic and hazardous environments.

IV. Experiments & Results

This section presents the results of the safe path planning using the optimized A* (OA*) algorithm, which integrates wildfire spread predictions from a Deep Learning model. The algorithm leverages binary fire masks to find safe evacuation routes while adhering to safety constraints, such as maintaining a buffer zone from fire-affected areas.

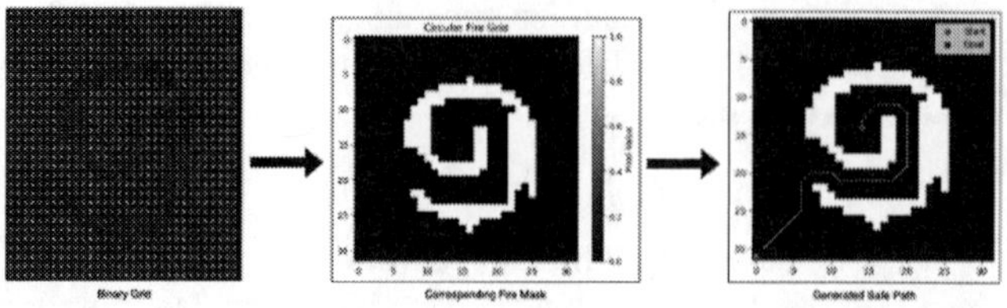

Figure 4: Application of OA* Algorithm on the Wildfire Spread Prediction results.

Fig. 4 builds upon Fig. 3, illustrating the application and outcome of the OA* algorithm on an input binary grid. The fire masks, generated from spread prediction outputs, are transformed into a binary grid to enable processing by the OA* algorithm. For clarity, these grids are further represented as fire maps. The final result depicts a red line, indicating the safe path between the start (green marker) and goal (blue marker), while adhering to all safety constraints.

A. Scenarios with Navigable Safe Routes

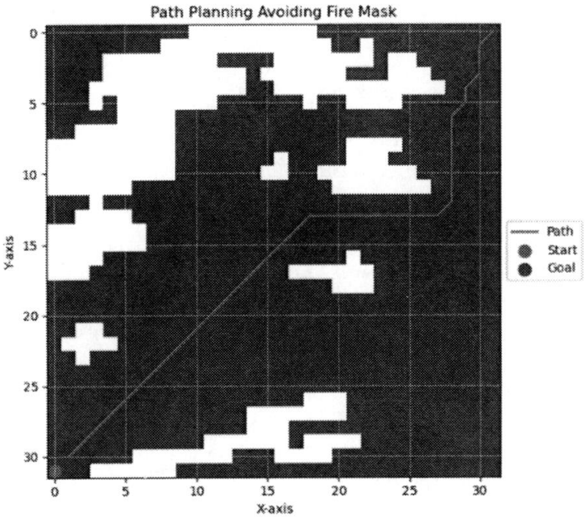

(a) Path generated in denser fire areas

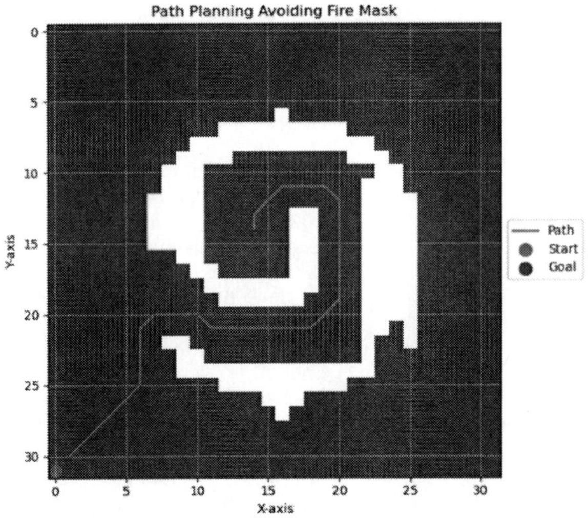

(b) Path generated in complex fire zones

Figure 5: Results of Navigable Path Based on Optimized A* Algorithm

The application of the OA* algorithm on wildfire spread predictions is demonstrated in Fig. 5, showing successful pathfinding from start to goal nodes, avoiding fire-affected regions. In these scenarios, the algorithm navigates a sparse region, selecting a direct route while respecting the safe buffer. Fig. 5a shows the algorithm can handle a denser fire area with narrow corridors, still ensuring safety by avoiding fire zones. The algorithm is equipped to find a longer, safer path around the fire in a large fire-affected area. Fig. 5b represents a complex spiral-shaped fire zone that is navigated successfully, demonstrating the algorithm's robustness.

The results demonstrate the OA* algorithm's ability to effectively utilize fire predictions to provide safe and efficient evacuation routes and validate that the OA* algorithm effectively balances safety and route efficiency. The implemented safety buffer ensures that the generated paths not only minimize distance but also maintain a safe distance from fire zones. The adaptability of the algorithm is evident, as it can navigate from sparse to highly complex fire scenarios without compromising safety.

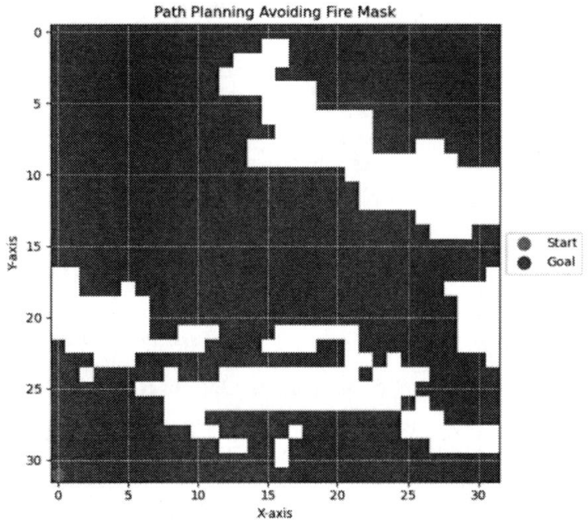

(a) Absence of safe distance from source to destination

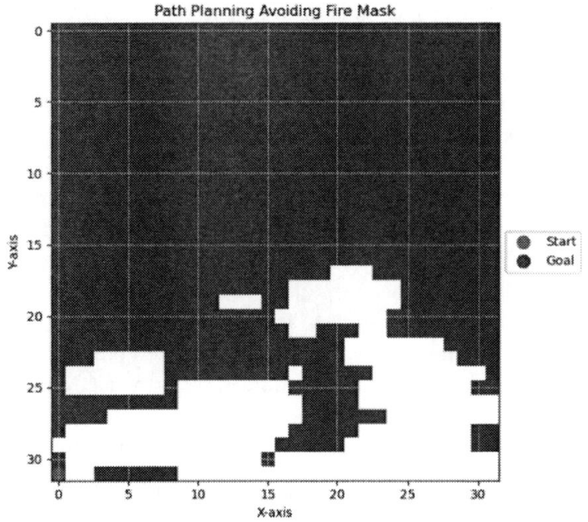

(b) Fire region overlaps or is too close to source/destination node.

Figure 6: Results of No Path Exist Based on Optimized A* Algorithm

B. Scenarios with No Safe Routes

In some cases, the OA* algorithm was unable to generate paths due to environmental constraints, as shown in Fig. 6. These scenarios highlight the algorithm's accuracy in recognizing situations where no safe route exists. Fig. 6a shows complete blockage of safe passages by fire zones and safety buffers resulting in no path being generated. Fig. 6b Fire regions overlap or are too close to the start or goal nodes, preventing any feasible path from being identified.

These results validate the robustness and reliability of the Optimized A* (OA*) algorithm in leveraging wildfire spread predictions for safe path planning. The algorithm consistently adheres to safety constraints, ensuring that only paths meeting safety standards are proposed while avoiding unsafe recommendations. Its ability to successfully navigate diverse fire scenarios demonstrates adaptability and effectiveness in providing safe evacuation routes. Simultaneously, the algorithm's precise recognition of no-path scenarios underscores its critical role in upholding safety protocols, making it a valuable tool for real-world applications like evacuation planning.

These outcomes validate the utility of the OA* algorithm as a practical tool for emergency planning and real-time decision-making, offering a promising solution for mitigating wildfire risks and enhancing evacuation strategies.

V. Conclusion & Future Work

In this work, we presented an Optimized A* (OA*) algorithm for safe path planning, leveraging wildfire spread predictions generated from deep learning models. The methodology integrates binary fire masks with a graph-based search algorithm to navigate through complex wildfire scenarios, ensuring safety while maintaining route efficiency. Experimental results demonstrate the algorithm's adaptability to varying fire patterns, from sparse to highly dense zones, and its ability to identify no-path scenarios when no safe evacuation route exists. This approach highlights the potential of integrating deep learning predictions with traditional search algorithms to address real-world challenges like emergency evacuation planning.

While the OA* algorithm has proven effective in ensuring safety and efficiency, future work can further enhance its capabilities. First, the quality of path planning outcomes is directly influenced by the accuracy and precision of wildfire spread predictions. Advances in deep learning techniques and the use of higher-resolution, more precisely delineated predictions can significantly improve the algorithm's performance. Additionally, incorporating dynamic updates to account for real-time changes in fire behavior and environmental conditions would enhance the adaptability of the system. Exploring alternative heuristic functions and multi-objective optimization techniques could also offer further improvements, allowing the algorithm to balance safety, travel time, and resource allocation more effectively.

References

[1] J. Halofsky, D. Peterson, and B. Harvey, "Changing wildfire, changing forests: The effects of climate change on fire regimes and vegetation in the pacific northwest, usa," *Fire Ecology*, vol. 16, p. 4, Dec. 2020. DOI: 10.1186/s42408-019-0062-8.

[2] J. D. Coop, S. A. Parks, C. S. Stevens-Rumann, S. M. Ritter, and C. M. Hoffman, "Extreme fire spread events and area burned under recent and future climate in the western usa," *Global Ecology and Biogeography*, vol. 31, no. 10, pp. 1949–1959, Apr. 2022, ISSN: 1466-8238. DOI: 10.1111/geb.13496.

[3] H. D. Safford, A. K. Paulson, Z. L. Steel, D. J. N. Young, and R. B. Wayman, "The 2020 california fire season: A year like no other, a return to the past or a harbinger of the future?" *Global Ecology and Biogeography*, vol. 31, no. 10, pp. 2005–2025, Apr. 2022, ISSN: 1466-8238. DOI: 10.1111/geb.13498.

[4] M. A. Moritz *et al.*, "Learning to coexist with wildfire," *Nature*, vol. 515, pp. 58–66, 2014.

[5] N. I. F. Center, *Annual wildfire summary report*, Report, 2021.

[6] P. L. Andrews, "The rothermel surface fire spread model and associated developments: A comprehensive explanation," 2018.

[7] P. L. Andrews, *Behave: Fire behavior prediction and fuel modeling system-burn subsystem*, US Department of Agriculture, Forest Service, Intermountain Research Station, 1986.

[8] M. A. Finney, *FARSITE: Fire Area Simulator-model development and evaluation*. 1998. DOI: 10.2737/rmrs-rp-4.

[9] I. Karafyllidis and A. Thanailakis, "A model for predicting forest fire spreading using cellular automata," *Ecological Modelling*, vol. 99, pp. 87–97, 1997.

[10] P. Cortez and A. Morais, "Using svm for forest fire prediction," *Journal of Environmental Management*, 2019.

[11] M. E. Green, K. Kaiser, and N. Shenton, "Modeling wildfire perimeter evolution using deep neural networks," *arXiv preprint arXiv:2009.03977*, 2020.

[12] F. Khennou and M. A. Akhloufi, "Improving wildland fire spread prediction using deep u-nets," *Science of Remote Sensing*, vol. 8, p. 100 101, 2023.

[13] S. Singla *et al.*, "Wildfiredb: An open-source dataset connecting wildfire occurrence with relevant determinants," 2021.

[14] S. Gerard, Y. Zhao, and J. Sullivan, "Wildfirespreadts: A dataset of multi-modal time series for wildfire spread prediction," *Advances in Neural Information Processing Systems*, vol. 36, pp. 74 515–74 529, 2023.

[15] E. DIJKSTRA, "A note on two problems in connexion with graphs.," *Numerische Mathematik*, vol. 1, pp. 269–271, 1959.

[16] P. E. Hart, N. J. Nilsson, and B. Raphael, "A formal basis for the heuristic determination of minimum cost paths," *IEEE transactions on Systems Science and Cybernetics*, vol. 4, no. 2, pp. 100–107, 1968.

[17] Z. Wang, S. Zlatanova, and P. van Oosterom, "Path planning for first responders in the presence of moving obstacles with uncertain boundaries," *IEEE Transactions on Intelligent Transportation Systems*, vol. 18, no. 8, pp. 2163–2173, 2017.

[18] Y. Xu, D. Li, H. Ma, R. Lin, and F. Zhang, "Modeling forest fire spread using machine learning-based cellular automata in a gis environment," *Forests*, vol. 13, no. 12, p. 1974, 2022.

Time-Series Topic Analysis of Large-Scale Social Media Data using Two-stage Clustering

Takako Hashimoto
Chiba University of Commerce
The University of Tokyo
Chiba/Tokyo, Japan
takako@cuc.ac.jp

Abstract—**Social media is a highly influential platform for sharing messages, photos, and videos. Understanding public perception through its vast data stream is essential. This study introduces a two-stage clustering method to extract coarse-grained topics from social media text data. First, graph clustering extracts micro-clusters from graphs generated based on the similarity of user posts, with each micro-cluster representing a fine-grained topic. The time series of these micro-clusters are then analyzed in the second stage through time series clustering to reveal coarse-grained topics. In this study, we consider applying this method to Yahoo! Japan News Comments related to the election of two specific candidates in Japan. This is expected to extract people's reactions to the candidates before and after the election.**

Keywords-social media analysis; knowledge discovery; graph mining; two-stage clustering; time series.

I. INTRODUCTION

Social media platforms, such as Yahoo! Japan News, are influential hubs for user interaction through messages and other media, such as photos and videos. Yahoo! Japan News, distributing around 7,500 daily articles and attracting 5 billion monthly visits, features an anonymous comment section where users can post opinions and react to others [1].

With the rapid growth of user-generated content, manually analyzing comments is impractical. To address this, we developed a two-stage clustering method to extract key topics and their temporal patterns from social media data [2]. This approach analyzes public perceptions, tracks opinion shifts, identifies misinformation, and evaluates communication effectiveness. It also bridges public sentiment with policymaking by providing actionable insights.

Unlike traditional topic modeling [3][4], which focuses on classification, our method combines graph clustering [5] of graphs based on the similarity of user posts with temporal analysis, enabling efficient processing of large-scale and sparse data while integrating topic and temporal evaluation.

This paper demonstrates the concept of our two-stage clustering method applying to various topical issues from Yahoo! Japan News Comments. In the first stage, we apply graph clustering to the word co-occurrence graph and extract micro-clusters corresponding to fine-grained topics of comments. We utilize Data Polishing algorithm [5][6] to achieve better scalability for data processing. We extract time series data for each micro cluster by counting the tweets posted within a defined time window. Specifically, we focus on 'burst' events, where a sudden spike in tweet activity corresponds to significant external events (such as election debates or breaking news). By examining these bursts, we can correlate the shifts in public perception with specific political or social occurrences. Finally, we apply time series clustering in the second stage to find the clusters corresponding to coarse-grained topics.

Our method has two key advantages: scalability and the ability to use both textual and temporal information. The method can handle large-scale social media data, making it suitable for long-term analysis. Additionally, by considering temporal patterns, we capture "bursty" activity [7] in the data, reflecting real-world events such as news and natural occurrences, which are essential for understanding underlying topics in social media discussions [8][9].

The rest of the paper is organized as follows. Section II surveys the related work. Our two-stage clustering method for discovering coarse-grained topics is briefly described in Section III. Next, Section IV explains our proposed method with a large-scale Yahoo! Japan News Comments data set regarding the election results of specific candidates in Japan. Finally, we summarize the results in Section VI.

II. RELATED WORK

Recent studies on social media data analysis have used Latent Dirichlet Allocation (LDA) [3] to identify topics in tweets [10][11]. However, LDA has limitations: it assumes documents contain multiple topics and require repeated word instances, making it unsuitable for social media posts that are generally short and low-quality. LDA also needs several hundred iterations, making it inefficient for large datasets, and it ignores temporal information such as timestamps [12][13][14]. Though extensions like X LDA [15] and dynamic LDA [16] address some of these issues, they still face other LDA limitations.

To overcome these problems, we propose a two-stage clustering algorithm using both word and timestamp data. This method applies graph clustering to identify micro-clusters (fine-grained topics) in social media data. Our proposed method, which applies graph clustering to identify fine-grained topics from social media data, extends previous work on topic modeling (e.g., LDA [16]). Unlike LDA, which requires repeated word occurrences in longer documents, our method is tailored to handle short, sparse data such as tweets. Furthermore, our approach incorporates time series analysis to capture the temporal dynamics often missing in conventional topic modeling techniques. Our clustering algorithm has five key characteristics: quantity, independence, coverage, granularity, and reproducibility. Existing methods like pattern mining [17], community mining [18], and DBscan [19] do not fully meet these criteria, but Data Polishing [5] offers a solution.

Event detection from social media streams is a popular research topic [20][21], with methods focusing on "bursty" events that lead to sharp rises in tweet activity. Our method, however, distinguishes between reactions to breaking news and general social media trends based on the similarity of temporal patterns.

Finally, while many studies focus on predicting social media trends [14][22][23], we develop an automatic method for discovering temporal patterns of collective human attention, helping to distinguish between external shocks and internal effects like word-of-mouth.

III. METHOD FOR DISCOVERING COARSE-GRAINED TOPICS FROM

We introduce a two-stage clustering method [2] to extract coarse-grained topics from social media data (Fig.1). The figure has the following four parts; A: Large-scale social media data (i.e., the tweets and timestamps). B: Graph defined by the similarity between user posts. C: Micro-clusters obtained by graph clustering (first stage clustering). D: Coarse-grained topic obtained by time series clustering (second stage clustering). In the C part, the gray circles represent a micro-cluster. Comments in a micro cluster share a fine-grained topic.

First, we construct a similarity graph of users' posts, where users' posts share similar words. For example, nodes (users' comments) are linked when sharing more than 50% of the words (Fig.1.A–B). Next, using a Data Polishing algorithm, graph clustering is applied to detect micro-clusters, representing fine-grained topics (Fig.1.C). Finally, time-series clustering groups these micro-clusters into coarse-grained topics, revealing how discussions evolve over time (Fig.1.D). This approach enhances scalability and integrates both textual and temporal features, making it suitable for analyzing large-scale comment datasets.

A. Graph Generation

We create the graph from users' posts, an undirected graph where each node represents a tweet/comment and an edge indicates tweet/comment similarity. A pair of tweets are connected if their Jaccard similarity coefficient[24] exceeds the threshold θE.

B. Graph Clustering

We identify fine-grained topics by clustering the users' posts graph to find micro-clusters (i.e., dense subgraphs) (Fig. 2). All the posts in a micro-cluster are expected to have a similar meaning.

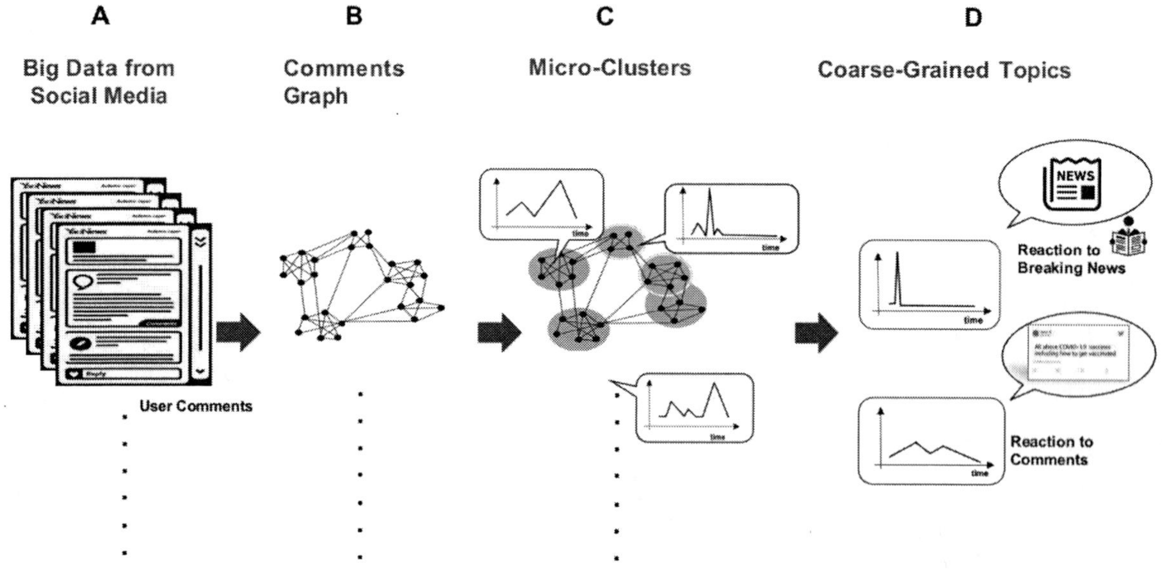

Figure 1. Proposed method (Two-stage clustering method) for discovering coarse-grained topics of public perceptions from social media data.

To identify fine-grained topics, we perform graph clustering to find micro-clusters (dense subgraphs) (Fig. 2). Users' posts within the same micro-cluster are highly similar. An edge is added between nodes (u, v) if the Jaccard similarity of their neighbor sets (N[u], N[v]) exceeds θ_{DP}. Non-satisfying edges are removed. Data Polishing iterates this process until the graph is divided into cliques, which are then identified as topics using maximal clique enumeration with MACE [25].

C. Time Series Clustering

While Data Polishing identifies topics, it often generates too many clusters for existing analysis. To reduce the number of clusters and improve interpretability, we use the users' posts timestamps (Fig.3). We divide time into windows and count the users' posts in each micro-cluster within those windows. Then, we apply K-Spectral Centroid (K-SC) [26] clustering to group micro-clusters with similar temporal patterns. K-SC is chosen for its robustness in capturing clusters using a similarity metric invariant to scaling and shifting, making it efficient for large datasets.

$$F = \sum_{j=1}^{K} \sum_{x_i \in C_j} \hat{d}(x_i, \mu_j)$$

where K is the number of clusters, xi is the *i-th* time series, and Cj is a set that represents the member of the *j-th* cluster. The K-SC's distance metric $\hat{d}(x, y)$ between the two time series (x and y) is defined as follows:

$$\hat{d}(x, y) = \min_{\alpha, q} \frac{\|x - \alpha y_q\|}{\|x\|}$$

where y_q is the result of shifting time series y by q time units, and $\| \ \|$ represent the l_2 norm.

IV. Discovering People's Perceptions about Specific Candidates before- or after- Elections

In this paper, we trying to apply the proposed method (Fig.1) to large-scale Yahoo! Japan News Comments to uncover public perceptions on coarse-grained topics. Yahoo! Japan News Comments, a widely used social media platform in Japan similar to X, often becomes a hot topic influencing public opinion on current issues.

A. Data

The dataset contains comments from Yahoo! Japan News mentioning the names of two candidates in local prefecture governor elections: hereafter CandidateA and CandidateB. For example, we suppose that CandidateA, initially a less known candidate, gained popularity during the election and finished as the runner-up. However, after the election, he faced significant criticism due to harassment allegations.

We can also suppose that CandidateB, the sitting governor of the Prefecture, faced harassment accusations, leading to his resignation. Despite the controversy, he gained support as the campaign progressed.

For these two candidates, Yahoo! Japan News that have one's name in the title will be the target data.

B. Adapting Two-stage Clustering

First, we segment each comment using the Japanese morphological analyzer MeCab [27] and remove stop words such as "kore" (this), "sore" (it), and "suru" (do). Next, we generate a user's comments text graph where each node represents a comment. Comments are connected if the Jaccard similarity coefficient of their word sets exceeds the threshold $\theta E = 0.3$, meaning the comments share more than 50% of words in common. In our two-stage clustering method, we set the threshold θ_{DP} to 0.2, and our experiments show that the results are robust to changes in this parameter.

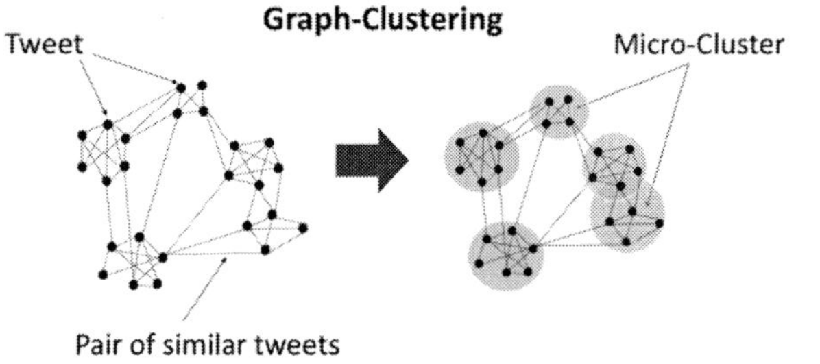

Figure 2. First stage clustering: Graph clustering.

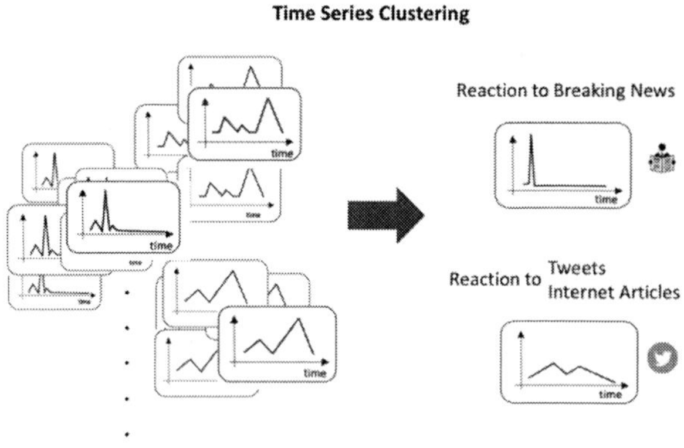

Figure 3. Second stage clustering: Time series clustering.

After identifying micro-clusters from the text graph, we extract coarse-grained topics by clustering the top 1,000 largest micro-clusters using time-series clustering (Sec. III C). The TimeSeriesKMeans algorithm was used, with the number of clusters K is set to 15, utilizing the Python package tslearn for time series analysis. We identified the cluster topics using ChatGPT.

C. *Analysis of Two-stage Clustering Result*

Fig.4 and Fig.5 present the two-stage clustering result examples (results of 2 clusters out of 15 clusters, respectively, as examples) for comments before and after the election of CandidateA and CandidateB, respectively. Fig.4-a and Fig. 5-a represent data before the election, while Fig.4-b and Fig. 5-b correspond to data after the election:

- Left (Word Cloud): Displays the most frequently used words, representing key themes in discussions
- Center (Time-Series Graph): Shows the temporal distribution of comments within each cluster, capturing shifts in public discourse.
- Right (Contents): Explains the cluster contents briefly.

This layout effectively visualizes how public sentiment and key discussion points changed before and after the election, helping to track emerging concerns and reactions. For example, we may be able to consider the following perspectives from people.

1) Before-election discussion on CandidateA: We could observe that preelection topics predominantly revolved around CandidateA's campaign strategies, political stance, qualifications, and media coverage. There was significant engagement with CandidateA's policies, leadership style, and future expectations, with public discourse centered on comparisons to other candidates and the feasibility of campaign promises. Media influence was also a recurring theme, though largely in the context of how it shaped CandidateA's perception among the public.

2) After-election discussion on CandidateA: We also obsereved that after election, the focus shifted from campaign strategies to critical evaluations of CandidateA's actions and credibility. The discussions became more emotionally charged, reflecting heightened public reactions to controversies, such as harassment allegations. Topics expanded to include broader concerns about political trustworthiness, media fairness, and societal trends. Discussions on gender biases and the humanity of politicians also emerged, reflecting a deeper exploration of societal and political issues. Additionally, the after-election discourse revealed increasing political distrust and dissatisfaction with the system.

3) Topic change from before-election to after-election for Candidate A: We can consider that the transition from before-election to after-election discourse reflects a shift from forward-looking campaign analysis to retrospective evaluations of political credibility, public trust, and societal implications, with a heightened emphasis on emotional and systemic critiques.

4) Before-election discussion on CandidateB: Prior to the election, the discourse primarily centered on CandidateBs campaign activities, leadership, political stance, and media coverage. Public opinion was shaped by discussions on CandidateB's qualifications as a governor, his policies, and his comparison to other political figures.

Media influence played a significant role in forming public perceptions, with many comments reflecting concerns

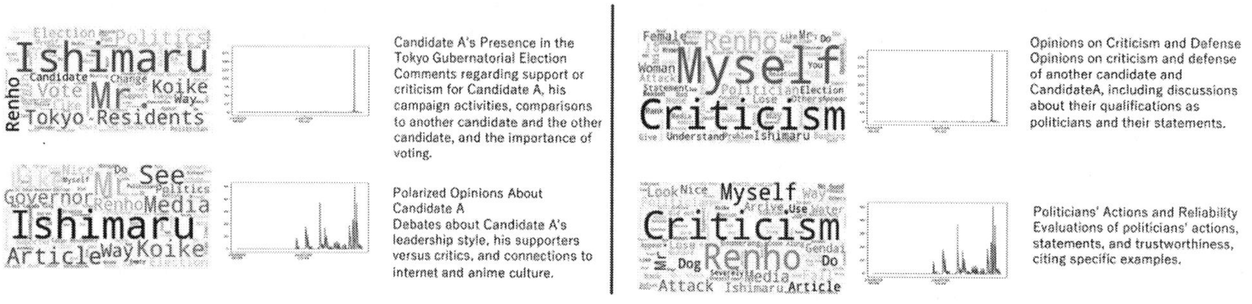

b. Cluster Examples of CandidateA - Before election b. Cluster Examples of CandidateA - After election

Figure 4. CandidateA's Coarse-grained topic examples obtained by two-stage clustering.

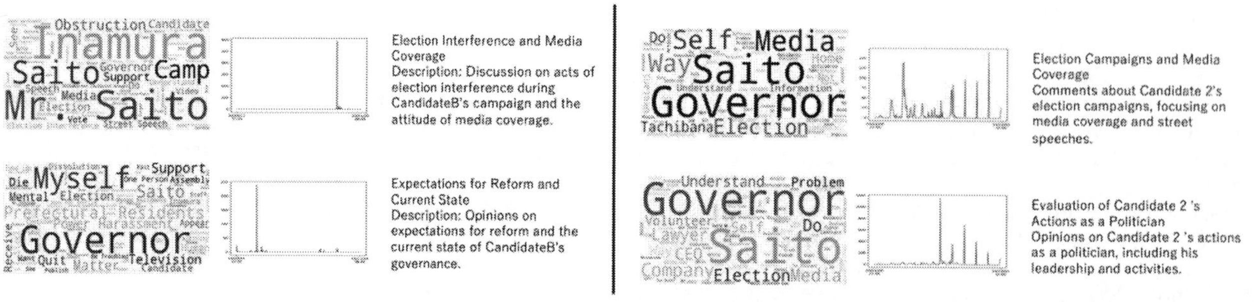

a. Cluster Examples of CandidateB - Before election b. Cluster Examples of CandidateB - After election

Figure 5. CandidateB's Coarse-grained topic examples obtained by two-stage clustering.

about the bias in reporting and CandidateBs reception in the media.

5) After-election discussion on CandidateB: After the election, the focus transitioned from before-election campaign strategies to more critical evaluations of CandidateB's actions and trustworthiness as a politician. The discourse became more polarized, with discussions increasingly reflecting public reactions to election results, controversies such as harassment allegations, and concern over political integrity. The after-election discussions also introduced themes of media fairness, misinformation, and political expectations, highlighting a growing dissatisfaction with politicians and the political system. There was a deeper exploration of topics such as trust in politicians, gender biases, and the social impact of political decisions.

6) Topic change from before-election to after-election for Candidate B: The shift from before-election to after-election discourse can be considered to illustrate a movement from campaign-focused discussions to more intense evaluations of political credibility, media influence, and societal concerns. The after-election phase may be marked by a heightened emphasis on trust issues, emotional responses, and systemic critiques.

This analysis is intended to indicate a direction, and the current data does not go far enough to capture this trend accurately. However, it is thought that it will become possible to grasp such trends by refining the data.

V. CONCLUSION

This study demonstrated the effectiveness of our two-stage clustering method in analyzing large-scale social media data, particularly in tracking public perceptions before and after elections. By applying the method to Yahoo! Japan News Comments, we identified key topics and their temporal evolution, uncovering shifts in political sentiment, media influence, and public trust.

Our findings highlight a notable shift in discussions moving from campaign strategies and candidate qualifications (before the election) to critical evaluations, controversies, and trust issues (after the election). Furthermore, increasing political polarization and concerns over media bias were observed, reflecting broader societal trends. Despite its effectiveness, limitations remain. Noisy data and evolving linguistic patterns could affect the method's performance. Future research will focus on enhancing robustness against noise and improving temporal clustering techniques. Beyond political analysis, this approach can also be adapted to analyze social media discussions around other societal issues, such as public health, environmental concerns, and economic policies. By applying this method to a variety of topics, we can gain deeper insights into how public opinion evolves in response to diverse challenges, enabling better-informed decision-making for policymakers, media analysts, and researchers.

ACKNOWLEDGMENT

This work was supported by JSPS KAKENHI Grant Number 23K21728.

REFERENCES

[1] Media personel blog, meditsubu (in japanese), https://media-radar.jp/contents/meditsubu/columns5-yahoo-news-pv/, Jan 2024, 2025.03.03.

[2] T. Hashimoto, T. Uno, Y. Takedomi, D. Shepard, M. Toyoda, N. Yoshinaga, M. Kitsuregawa, and R. Kobayashi, "Two-stage clustering method for discovering people's perceptions: A case study of the covid-19 vaccine from twitter," 2021 IEEE International Conference on Big Data (Big Data), pp. 614–621, 2021.

[3] D. M. Blei, A. Y Ng, and M. I. Jordan, "Latent dirichlet allocation," Journal of machine Learning research, 3: pp. 993–1022, 2003.

[4] D. M. Blei, "Probabilistic topic models," Communications of the ACM, 55(4): pp. 77–84, 2012.

[5] T. Uno, H. Maegawa, T. Nakahara, Y. Hamuro, R. Yoshinaka, and M. Tatsuta, "Micro-clustering by data polishing," 2017 IEEE International Conference on Big Data (Big Data), pp. 1012–1018. IEEE, 2017.

[6] T. Hashimoto, D. L. Shepard, T. Kuboyama, K. Shin, R. Kobayashi, and T. Uno, "Analyzing temporal patterns of topic diversity using graph clustering," The Journal of Supercomputing, 77(5): pp. 4375–4388, 2021.

[7] J. Kleinberg, "Bursty and hierarchical structure in streams," Data mining and knowledge discovery, 7(4): pp. 373–397, 2003.

[8] J. Lehmann, B. Gonçalves, J. J. Ramasco, and C. Cattuto, "Dynamical classes of collective attention in twitter," Proceedings of the 21st international conference on World Wide Web, pp. 251–260, 2012.

[9] R. Kobayashi, P. Gildersleve, T. Uno, and R. Lambiotte, "Modeling collective anticipation and response on wikipedia," Proceedings of the International AAAI Conference on Web and Social Media, 15(1): pp. 315–326, 2021.

[10] R. J. Medford, S. N Saleh, A. Sumarsono, T. M Perl, and C. U. Lehmann, "An "infodemic": leveraging high-volume twitter data to understand early public sentiment for the coronavirus disease 2019 outbreak," Open Forum Infectious Diseases, 7(7):ofaa258, 2020.

[11] J. C. Lyu, E. L. Han, and G. K Luli, "Covid-19 vaccine–related discussion on twitter: Topic modeling and sentiment analysis," Journal of medical Internet research, 23(6):e24435, 2021.

[12] J. Hurlock and M. L Wilson, "Searching twitter: Separating the tweet from the chaff," Fifth International AAAI Conference on Weblogs and Social Media, pp. 161–168, 2011.

[13] C. Castillo, M. Mendoza, and B. Poblete, "Information credibility on twitter," Proceedings of the 20th international conference on World wide web, pp. 675–684, 2011.

[14] T. Murayama, S. Wakamiya, E. Aramaki, and R. Kobayashi, "Modeling the spread of fake news on twitter," Plos one, 16(4):e0250419, 2021.

[15] W. Xin Zhao, J. Jiang, J. Weng, J. He, E. P. Lim, H. Yan, and X. Li, "Comparing twitter and traditional media using topic models," European conference on information retrieval, pp. 338–349. Springer, 2011.

[16] D. M. Blei and J. D. Lafferty, "Dynamic topic models," Proceedings of the 23rd international conference on Machine learning, pp. 113–120, 2006.

[17] T. Uno, et al., "Lcm ver. 2: Efficient mining algorithms for frequent/closed/maximal itemsets," Fimi, volume 126, 2004.

[18] R. Kumar, P. Raghavan, S. Rajagopalan, and A. Tomkins. Trawling the web for emerging cybercommunities. Computer networks, 31(11-16): pp. 1481–1493, 1999.

[19] M. Ester, et al., "A density-based algorithm for discovering clusters in large spatial databases with noise," kdd, volume 96, pp. 226–231, 1996.

[20] M. Hasan, M. A. Orgun, and R. Schwitter, "A survey on real-time event detection from the twitter data stream," Journal of Information Science, 44(4): pp. 443–463, 2018.

[21] C. Comito, A. Forestiero, and C. Pizzuti, "Bursty event detection in twitter streams," ACM Transactions on Knowledge Discovery from Data (TKDD), 13(4): pp. 1–28, 2019.

[22] R. Kobayashi and R. Lambiottem "Tideh: Time-dependent Hawkes process for predicting retweet dynamics," Tenth International AAAI Conference on Web and Social Media, 2016.

[23] J. Proskurnia, P. Grabowicz, R. Kobayashi, C. Castillo, P. C. Mauroux, and K. Aberer, "Predicting the success of online petitions leveraging multidimensional time-series," Proceedings of the 26th International Conference on World Wide Web, pp. 755–764, 2017.

[24] P. Jaccard, "The distribution of the flora in the alpine zone," 1. New phytologist, 11(2): pp. 37–50, 1912.

[25] K. Makino and T. Uno, "New algorithms for enumerating all maximal cliques," Scandinavian workshop on algorithm theory, pp. 260–272. Springer, 2004.

[26] J. Yang and J. Leskovec, "Patterns of temporal variation in online media," Proceedings of the fourth ACM international conference on Web Search and Data Mining, pp. 177–186, 2011.

[27] T. Kudo, "Mecab: Yet another part-of-speech and morphological analyzer," https://sourceforge.net/projects/mecab/, 2013, 2025.03.03

Evaluating the Potential of SHAP-Based Feature Selection for Improving Classification Performance

Ashis Kumar Mandal
Department of Computer Science and Engineering
Hajee Mohammad Danesh Science and Technology University
Dinajpur, Bangladesh
e-mail:ashis@hstu.ac.bd

Basabi Chakraborty
Madanapalle Institute of Technology and Science
Andhra Pradesh, India
Iwate Prefectural University
Iwate, Japan
email:basabi@iwate-pu.ac.jp

***Abstract*—Feature selection is an important preprocessing step in developing efficient and accurate classification models. Among various techniques, recently SHapley Additive exPlanations (SHAP)-based feature selection has gained attention for its interpretability and ability to quantify the contributions of individual features to model predictions. This study investigates the effectiveness of SHAP-based feature selection technique, specifically focusing on Linear SHAP, in improving classification performance. The research utilizes 10 diverse datasets to evaluate Linear SHAP's capability in identifying relevant features for classification tasks. The performance of Linear SHAP is assessed across varying percentages of selected features and compared to classification models without feature selection. Three popular filter-based feature selection approaches: Chi-square(Chi^2), Mutual Information, and Correlation-based methods are also used for feature selection with the same bench mark data sets. Comparative analysis, supported by statistical significance tests, demonstrates that Linear SHAP performs equally well to the traditional methods while offering the added benefit of interpretability. The findings suggest that Linear SHAP is a viable and promising alternative to established feature selection techniques in the realm of classification tasks.**

Keywords-Feature Selection; SHapley Additive exPlanations (SHAP); classification models; machine learning.

I. INTRODUCTION

Feature selection plays a pivotal role in machine learning models, particularly in classification tasks, by identifying and retaining the most relevant features from high-dimensional datasets in order to improve performance of the model. This process not only enhances model accuracy but also improves computational efficiency, reduces overfitting, and increases interpretability [1]. While traditional feature selection methods, including filter [2], wrapper [3], and embedded approaches [4], have been widely studied and applied to real world problems, they often struggle to find out efficient and optimal feature subset from high-dimensional datasets and sometime may not able to capture complex feature to feature interactions.

In recent years, eXplainable Artificial Intelligence (XAI) has gained significant attention in the development of trustworthy AI systems for recommendation and decision-making, particularly in high-risk application areas such as healthcare, finance, and control. As feature selection is an important preprocessing step, interpretable feature selection leads to the improvement of explanation ability of any pattern recognition or machine learning model. SHapley Additive exPlanations (SHAP) [5] has recently gained attention as an interpretable framework for understanding the contributions of individual features in machine learning models. While SHAP is widely recognized as a tool for model explanation, its utility as a feature selection mechanism is less explored. Linear SHAP [6], a variant designed for linear models, offers computational efficiency and interpretability, making it a promising candidate for feature selection. Its ability to quantify feature importance in an additive and consistent manner provides a unique advantage for understanding the relationship between features and predictions. However, a comprehensive evaluation of SHAP-based feature selection before classification tasks and its comparison with established methods is still lacking in the literature.

This paper seeks to fill the existing gap by thoroughly investigating the potential of SHAP-based feature selection to enhance classification performance. We outline a systematic approach to evaluate SHAP-based feature selection in comparison to traditional methods, focusing on the following key questions:

1) How does SHAP-based feature selection perform relative to approaches that do not utilize feature selection in terms of classification performance?
2) What is the impact of reducing the number of features on classification performance?
3) How does the performance of SHAP-based feature selection compare with popular filter-based feature selection methods?

To address the above questions, we have utilized a diverse array of benchmark datasets. Our methodology have involved a rigorous comparison of SHAP-based feature selection against well-established techniques, such as Chi-square, Mutual Information, and Correlation-based feature selection approaches.

The rest of this paper is organized as follows: Section II provides a theoretical background on feature selection in general and SHAP based approaches for feature selection. Section III describes our proposed methodology for evaluating SHAP-based feature selection in comparison to other existing state of the art approaches. Section IV presents the experimental results followed by a short section on discussion on the limitations of this study. Finally, Section VI concludes the paper and outlines directions for future research.

II. Theoretical Background and Related Study

A. Feature selection

Feature selection aims to identify the most informative and discriminative features while eliminating irrelevant or redundant ones. Based on their interaction with the learning model, feature selection methods are categorized as filter, wrapper, and embedded approaches. Filter methods evaluate the relevance of features using statistical measures or intrinsic properties of the data. The most common and widely used techniques include Chi-square, Correlation-based analysis, Mutual Information and ANOVA [7] [8]. Wrapper methods involve the classifier model for evaluation of the feature subsets by training and validating the model on the data, with examples such as, Recursive Feature Elimination (RFE) and forward or backward selection [9]. Embedded methods integrate feature selection directly into the model training process, as seen with L1 regularization (Lasso) [10]. Feature selection can also be classified based on its approaches, such as feature ranking, where the top-K features are selected, or feature subset selection, which aims to identify an optimal or near-optimal subset of features [11]. These methods collectively improve model performance, reduce dimensionality, and enhance interpretability of the classification task as a whole.

B. SHAP

The idea of Shapley values originated from cooperative game theory. In game theory, the Shapley value of a player is the average marginal contribution of the player in a cooperative game. The Shapley value framework was developed by Llyod Shapley [12] which is based on some fairness axioms. This framework fairly allocates a contribution score to each player, reflecting their role in achieving the overall payoff. Lundberg applied the idea to machine learning, in which SHAP(SHapley Additive exPlanations) treats each feature as a player and calculates its contribution to the model's predictions [13]. SHAP approximates Shapley value by computing the contribution to a model's prediction of every subsets of features, given a dataset with m features. In this context, this approach offers a reliable and interpretable means of assessing the influence of individual feature or a subset of features on model outputs, facilitating both local and global insights into the model's behavior [14]. Computation of exact solution of Shapley values is quite infeasible for large number of inputs (players or features) due to the exponential nature of the problem. SHAP approximate the solutions through special weighted linear regression for any model or throughout different assumptions about feature dependence for ensemble tree models [15].

C. SHAP-Based Feature Selection Approaches

To date, several research efforts have utilized SHAP to improve model interpretability and examined its application in feature selection [16]. SHAP has been used effectively in medical diagnostics to improve the interpretability of the model. Huang et al. [17] developed a logistic regression model for the detection of heart failure, integrating SHAP for the global interpretation of the significance of features. This approach demonstrated superior precision compared to traditional methods by focusing on clinically relevant features. Luo et al. utilized SHAP to predict water quality indices, illustrating its capacity to highlight the most influential features in hydrological datasets [18]. Gehlot et al. [19] employed SHAP to enhance the explainability of machine learning models for surface electromyography-based hand gesture recognition. This study integrated SHAP scores to refine feature subsets, increasing the precision and interpretability of the model. SHAP has proven valuable in cancer detection, as demonstrated by a study that combined SHAP with machine learning for metabolomic analysis in breast cancer patients. This hybrid approach outperformed traditional selection methods, providing detailed insights into feature contributions [20]. In the domain of NLP, Ramanujam et al. [21] applied SHAP to select features for classifying spam SMS in Dravidian languages. Santos et al. [22] explored SHAP for efficient feature selection in the domain of industrial fault diagnosis. Here, an explainable artificial intelligence (XAI) technique is incorporated to meticulously select optimal features for the machine learning (ML) models. The chosen ML technique for the tasks of fault detection, classification, and severity estimation is the support vector machine (SVM). The interpretable analysis method based on Shapley values effectively enhances the performance of recognizing similar gestures and provides valuable insights into the decision-making process of recognition models in the research work of Wang et.al [23]. Overall, the motivation behind the use of shapley value or SHAP in selection of optimal features for a classification task is to build interpretable model capable of explaining the behavior of the model in the decision process.

III. Methodology

The primary objective of this research is to perform experiments using features identified by a SHAP-based approach to build an efficient classifier model. To achieve this, we collect 10 datasets for experimentation. The overall workflow of the task is illustrated in Figure 1. Initially, the datasets are preprocessed, which involves data cleaning, imputation of missing values, date encoding, normalization, and data balancing. After necessary processing, the datasets are partitioned into training and testing sets. The training datasets are then used for feature selection via the SHAP-based approach. A machine learning model is built using the selected features from the training datasets. The testing datasets are used to evaluate the performance of the models. Experiments are conducted with varying percentages of selected features for each dataset to analyze the effect of feature subset size on classifier performance. For comparison purposes, CHI2, Mutual Information, and Correlation-based feature selection methods are also employed for feature ranking.

A. Datasets

For our experimental analysis, we selected ten diverse datasets from the UCI Machine Learning Repository [24]. These datasets vary in their number of features, instances,

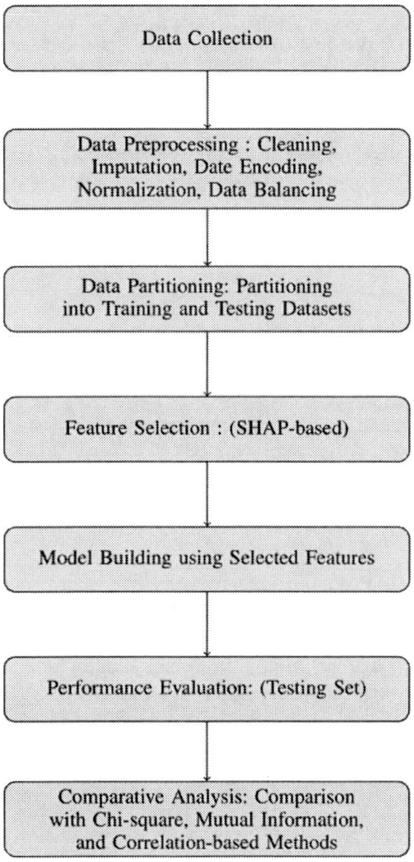

Figure 1. Workflow of the Research Methodology.

and classes, providing a comprehensive test bed for our experiments. Table I presents an overview of these datasets.

TABLE I. DATASETS

Datasets	No. of feature	No. of instances	No. of classes
Breast-w	9	683	2
Clean	166	476	2
Hepatitis	19	155	2
Parkinsons	22	195	2
Promoters	57	107	2
Qsar-biodeg	42	1055	2
Sonar	60	208	2
Spect	23	267	2
Spectf	45	349	2
Wisconsin	17	110	7

B. Data Preprocessing

We cleaned the datasets to remove inconsistent entries and address missing values. For continuous features, we imputed missing values using the mean, while for discrete features, we used the median. To process categorical data, we applied one-hot encoding, which transforms categorical variables into binary vectors by creating separate binary columns for each category. A value of 1 indicates the presence of a specific category, while 0 indicates its absence. Additionally, we used label encoding to convert each category into a unique integer, enabling machine learning models to process categorical data as numeric inputs.

To address class imbalances and reduce overfitting, we performed data balancing using the Synthetic Minority Oversampling Technique (SMOTE), which generates synthetic samples for underrepresented classes.

Finally, we normalized the features to ensure they were represented on a uniform scale. We applied mini-max normalization to scale all feature values to a range between 0 and 1, preventing features with larger magnitudes from dominating the model training process.

C. Data Partitions and feature selection

After preprocessing, each dataset was partitioned into training and testing sets, with 80% of the data allocated for training and 20% for testing. The training dataset was primarily utilized for feature selection.

For feature selection, we employed a SHAP-based approach, specifically using the SHAP Linear Explainer to compute feature importance. Logistic Regression was selected as the underlying model for this process. The Linear Explainer was chosen due to its ability to calculate feature contributions with minimal computational overhead compared to kernel-based or tree-based SHAP methods. This approach provided a quantitative measure of feature importance, enabling us to rank the features and select the top n% for further analysis.

To compare SHAP-based feature selection with other methods, we employed three popular rank-based feature selection techniques: Chi-square (Chi^2), mutual information, and Correlation-based methods. Each of these methods was applied to the same dataset, and the top n% features were selected for each case.

For each feature selection approach, we evaluated the performance of machine learning models trained on the selected features. Logistic Regression was chosen as the classification model for this evaluation. Using the training dataset, we built models based on the features selected by each feature selection method. The classification accuracy of these models was then assessed using the testing dataset to evaluate the effectiveness of the respective feature selection approaches. To assess the impact of feature selection, we compared the models' performance with 25%, 50%, 75%, and 100% of the features, where 100% represents the dataset without any feature selection.

D. Experimental setup

For each feature selection approach, the experiment was conducted five times using different seed values to ensure robust results. The average classification accuracy across these runs was calculated to evaluate the performance of the models. The implementation was developed in Python. For SHAP-based feature selection, the SHAP library was utilized, while the scikit-learn library was used for building and evaluating machine learning models. The entire experiment was performed on a system with an Intel Core i5 processor, 8 GB of RAM, and running the Windows operating system.

IV. Experimental Results and Analysis

In this section, we present the results obtained using SHAP-based feature selection approaches and compare them with three other ranking-based filter type feature selection approaches.

Table II illustrates the average classification accuracy over 10 independent runs. It presents the results of the linear SHAP-based feature selection approach across different selection percentages. The analysis involves 10 datasets, considering feature selection percentages of 25%, 50%, 75%, and 100% (i.e., without feature selection).

When comparing feature selection with no feature selection, it is observed that reducing the features to 50% or 75% yields an average classification accuracy similar to that achieved without applying feature selection. Specifically, for 75% selected features using the SHAP-based approach, eight datasets exhibit the same average classification accuracy compared to the results obtained without feature selection. In one case, the SHAP-based approach outperforms, while in another case, it performs slightly worse compared to the scenario without feature selection.

Similarly, when 50% of the features are selected using the SHAP-based approach, four datasets demonstrate improved average classification accuracy compared to the results without feature selection. One case shows significantly better performance, while the remaining datasets show slightly lower performance compared to the case of without feature selection.

Table III. compares the performance of Linear SHAP with three filter-based feature selection methods: Chi2, Mutual Information, and Correlation. In this comparison, the average classification accuracy is considered with 50% of the features selected. It is found that Linear SHAP consistently demonstrates superior performance across various datasets, emphasizing its effectiveness in feature selection tasks. Specifically, it outperforms Chi2 in datasets such as Qsar-biodeg, Wisconsin, and Clean, while achieving equal performance in Breast-w, Parkinsons, Sonar and Spectf. Compared to Mutual Information, Linear SHAP shows notable advantages in Wisconsin and Qsar-biodeg, although Mutual Information performs better in Parkinsons and Promoters, with both methods yielding similar results in Breast-w. When evaluated against the Correlation method, Linear SHAP exhibits strengths in Wisconsin, Clean and Qsar-biodeg while Correlation is more effective in Parkinsons, with identical outcomes observed for Breast-w. Overall,

TABLE II. Performance Analysis of Linear SHAP-Based Feature Selection Based on Average Classification Accuracy Across Different Selection Percentages

Datasets	Percentage of selected feature			
	25%	50%	75%	100% (Without Feature Selection)
Breast-w	0.95	0.97	0.97	0.97
Clean	0.77	0.81	0.81	0.81
Hepatitis	0.71	0.76	0.78	0.78
Parkinson	0.74	0.76	0.76	0.76
Promoters	0.75	0.76	0.76	0.76
Qsar-biodeg	0.80	0.82	0.83	0.83
Sonar	0.76	0.75	0.78	0.78
Spect	0.71	0.69	0.69	0.68
Spectf	0.79	0.78	0.79	0.79
Wisconsin	0.95	0.96	0.96	0.97

Linear SHAP demonstrates better performance compared to the other methods in three out of 10 data sets, while maintains equal performance in six out of 10 data sets. Only in case of Parkinsons dataset, Mutual Information and Correlation based methods produce better results than Linear SHAP. From these findings, it can be inferred that Linear SHAP demonstrates robust and consistent performance, establishing itself as a reliable approach for feature selection across diverse datasets.

TABLE III. Accuracy Comparison of Linear SHAP Across Ranking-Based Feature Selection Approaches

Datasets	Linear Shap	Chi^2	Mutual Information	Correlation
Breast-w	0.97	0.97	0.97	0.97
Clean	0.81	0.78	0.77	0.78
Hepatitis	0.76	0.77	0.77	0.77
Parkinsons	0.76	0.76	0.78	0.80
Promoters	0.76	0.78	0.79	0.78
Qsar-biodeg	0.82	0.77	0.80	0.80
Sonar	0.75	0.75	0.76	0.74
Spect	0.69	0.70	0.70	0.70
Spectf	0.78	0.78	0.77	0.78
Wisconsin	0.96	0.93	0.93	0.93

TABLE IV. Pairwise Wilcoxon Signed-Rank Test p-values for Linear SHAP Compared to Other Methods

Method	Chi2	Mutual Info	Correlation
Linear SHAP	0.40	0.95	0.85

Table IV shows a statistical comparison of Linear SHAP with three different approaches using a pairwise Wilcoxon

Signed Rank test. All three comparisons show relatively high p-values (< 0.05), suggesting that Linear SHAP's performance is not statistically different from any of the other three approaches. This implies that Linear SHAP performs comparably similar to the alternative methods in this analysis.

V. Discussion and Limitation

The experimental results and their analysis has been presented in the previous section. In this study, we have used linear models of classification and Linear SHAP version of SHAP implementation is chosen for feature selection as it is computationally less expensive. The primary objective is to explore the viability of SHAP-based methods for rank based feature selection in comparison to traditional feature selection approaches: Chi2, Mutual Information, and Correlation-based methods. Linear SHAP is utilized to select features, with its performance evaluated using linear machine learning models across different percentages of selected features. The subset based feature selection methods are not examined in this work. Though we have not presented detail results of computational cost involved in feature selection methods, it seems that the computational cost of Linear SHAP is comparable with other popular optimal feature selection algorithms for datasets having moderate number of features.

Experimental results demonstrate that Linear SHAP is an effective tool for feature selection, consistently identifying relevant features and maintaining competitive performance across diverse datasets. The comparative analysis highlights its strengths and reliability when compared with Chi2, Mutual Information, and Correlation-based approaches. Statistical significance tests indicate that Linear SHAP performs equivalently to these traditional methods in several cases while offering the added advantage of interpretability, making it particularly valuable for applications where interpretability is a priority.

VI. Conclusion

In this work, a preliminary study has been done to assess the potential of SHAP- based feature selection method in comparison with other popular filter based methods with a limited number of bench mark data sets of moderate dimension. The results are encouraging. It shows that the performance of SHAP based method is comparable to other popular rank based feature selection algorithms. At present, the explanation capability of a decision model is very much valued in many practical applications in the area of medical or finance. In this context, SHAP based methods have a solid background and mathematical foundation to facilitate interpretability of the selected features. The development of effective SHAP based optimal feature selection algorithm can have a great impact in designing explainable decision systems.

As this study is limited to rank based feature selection algorithms involving linear SHAP, detail experiments with more sophisticated versions of SHAP implementations with more high dimensional data sets are needed for proper evaluation of SHAP based methods, especially regarding computational cost. Future study could also focus on developing hybrid methods that combine Linear SHAP with other feature selection techniques to leverage complementary strengths. Additionally, evaluating the scalability and generalization capability of these methods on more complex datasets and classification tasks would provide deeper insights into their broader applicability for interpretable feature selection.

Acknowledgment

This research project was supported by Japan Society of Promotion of Science (JSPS) KAKENHI Grant Number JP 24K15089 Type: Kaken C.

References

[1] P. Dhal and C. Azad, "A comprehensive survey on feature selection in the various fields of machine learning," *Applied Intelligence*, vol. 52, no. 4, pp. 4543–4581, 2022, ISSN: 1573-7497. DOI: 10.1007/s10489-021-02550-9.

[2] A. Bommert, T. Welchowski, M. Schmid, and J. Rahnenführer, "Benchmark of filter methods for feature selection in high-dimensional gene expression survival data," *Briefings in Bioinformatics*, vol. 23, no. 1, pp. 1–13, 2022, ISSN: 1477-4054. DOI: 10.1093/bib/bbab354. eprint: https://academic.oup.com/bib/article-pdf/23/1/bbab354/42229629/bbab354.pdf.

[3] J. Maldonado, M. C. Riff, and B. Neveu, "A review of recent approaches on wrapper feature selection for intrusion detection," *Expert Systems with Applications*, vol. 198, p. 116 822, 2022, ISSN: 0957-4174. DOI: https://doi.org/10.1016/j.eswa.2022.116822.

[4] N. Mahendran and D. R. V. P M, "A deep learning framework with an embedded-based feature selection approach for the early detection of the alzheimer's disease," *Computers in Biology and Medicine*, vol. 141, p. 105 056, 2022, ISSN: 0010-4825. DOI: https://doi.org/10.1016/j.compbiomed.2021.105056.

[5] S. M. Lundberg and S.-I. Lee, *Consistent feature attribution for tree ensembles*, 2018. arXiv: 1706.06060 [cs.AI].

[6] L. Schulte, B. Ledel, and S. Herbold, "Studying the explanations for the automated prediction of bug and non-bug issues using lime and shap," *Empirical Software Engineering*, vol. 29, no. 4, p. 93, 2024, ISSN: 1573-7616. DOI: 10.1007/s10664-024-10469-1.

[7] I. Guyon and A. Elisseeff, "An introduction to variable and feature selection," *Journal of Machine Learning Research*, vol. 3, pp. 1157–1182, 2003.

[8] H. Liu and L. Yu, "Toward integrating feature selection algorithms for classification and clustering," *IEEE Transactions on Knowledge and Data Engineering*, vol. 17, no. 4, pp. 491–502, 2005.

[9] R. Kohavi and G. H. John, "Wrappers for feature subset selection," *Artificial Intelligence*, vol. 97, no. 1-2, pp. 273–324, 1997.

[10] R. Tibshirani, "Regression shrinkage and selection via the lasso," *Journal of the Royal Statistical Society: Series B (Methodological)*, vol. 58, no. 1, pp. 267–288, 1996.

[11] Y. Saeys, I. Inza, and P. Larrañaga, "A review of feature selection techniques in bioinformatics," *Bioinformatics*, vol. 23, no. 19, pp. 2507–2517, 2007.

[12] L. S. Shapley, "A value for n-person games," *Contributions to the Theory of Games*, vol. 2, no. 28, pp. 307–317, 1953.

[13] S. M. Lundberg and S.-I. Lee, "A unified approach to interpreting model predictions," *Advances in Neural Information Processing Systems*, vol. 30, pp. 4765–4774, 2017.

[14] C. Molnar, *Interpretable machine learning: A guide for making black box models explainable*. Lulu. com, 2020.

[15] S. Lundberg, G. Erion, and H. e. a. Chen, "From local explanations to global understanding with explainable ai for trees," *Nat Mach Intell*, pp. 56–67, 2020. DOI: https://doi.org/10.1038/s42256-019-0138-9.

[16] E. Wilson and D. M. Eler, "From explanations to feature selection: Assessing shap values as feature selection mechanism," *Proc.33rd IEEE SIBGRAPH Conference on Graphics, Patterns and Images*, pp. 340–346, 2020.

[17] H. Huang, J. Guan, and C. Feng, "Fluid volume status detection model for patients with heart failure based on machine learning methods," *Heliyon*, vol. 11, no. 1, e41127, 2025.

[18] H. Luo, C. Xiang, and L. Zeng, "Shap based predictive modeling for water quality index calculation in hydrological studies," *Scientific Reports*, 2024.

[19] N. Gehlot, A. Jena, and A. Vijayvargiya, "Surface electromyography based explainable ai fusion framework for hand gesture recognition," *Engineering Applications of Artificial Intelligence*, vol. 137, pp. 109–119, 2024, ISSN: 0952-1976. DOI: https://doi.org/10.1016/j.engappai.2024.109119.

[20] F. Yagin and Y. Gormez, "Hybrid methodology integrating shap for metabolomic analysis in breast cancer patients," *Frontiers in Oncology*, 2024.

[21] E. Ramanujam, K. Thirumalai, and A. Abirami, "Fsshap: Global interpretable feature selection using xai for the classification of spam sms in dravidian languages," *IEEE MultiMedia*, pp. 1–11, 2024. DOI: 10.1109/MMUL.2024.3508765..

[22] M. L. Santos, A. Guedis, and I. S. Gendris, "Shapley additive explanations (shap) for efficient feature selection in rolling bearing fault diagnosis," *Machine Learning and Knowledge Extraction*, vol. 6, no. 1, pp. 316–341, 2024.

[23] F. Wang, X. Ao, M. Wu, S. Kawata, and J. She, "Explainable deep learning for semg-based similar gesture recognition: A shapley-value-based solution," *Information Sciences*, vol. 672, p. 120667, 2024, ISSN: 0020-0255. DOI: https://doi.org/10.1016/j.ins.2024.120667.

[24] M. Kelly, R. Longjohn, and K. Nottingham, *The UCI machine learning repository, https://archive.ics.uci.edu.*

Visualizing Proximity of Audio Signals from Different Musical Instruments - A Two Step Approach

Goutam Chakraborty *
Madanapalle Institute Of Technology & Science, Madanapalle, India
Iwate Prefectural University, Information Science, IPU, Iwate, Japan
Email: goutam@iwate-pu.ac.jp

Cedric Bornand†
University of Applied Sciences HES-SO
Yverdon-les-Bains, Switzerland
Email: cedric.bornand@heig-vd.ch

A. Lokesh‡, Subhash Molaka§, Praveen Kumar Reddy Sangati¶, Lakshman Patti‖
Department of Computer Science Engineering and Artificial Intelligence‡§¶‖
Madanapalle Institute Of Technology & Science, Madanapalle, India‡§¶‖
Email: lokeshreddy2680@gmail.com‡, molakasubhash@gmail.com§,
prawinreddy1909@gmail.com¶, lakshmanpatti99@gmail.com‖

***Abstract*—We perceive music from various perspectives - the melody, the rhythm, the emotions or passions they evoke, the richness of sound, and how it correlates with the time of the day (like Morning Raga) or with seasons (like Vivaldi's Four Seasons). This is a multimodal classification challenge for which correct data annotation is a difficult issue. In this work, we propose a method for visualizing audio signals from various musical instruments to identify their variances and quantify their similarities and distances. The appropriate tools (algorithms) for this task were identified by experimental analysis. The work is conducted in two stages: the first is audio feature extraction and compression, and the second is the projection of high-dimensional audio features on a two-dimensional plane using various unsupervised visualization techniques. The aim is to determine which feature compression and visualization tools can produce clearly separated clusters of audio signals. The features of the STFT spectrogram extracted using CNN provide the best compressed representations, which are better visualized using t-SNE and UMAP techniques, achieving silhouette scores of 84% and 81%, respectively. The STFT spectrogram features are compressed more effectively using UNet, resulting in improved cluster visualization with t-SNE, UMAP, and even with PCA, with silhouette scores of around 75%.**

Keywords*- *MFCC; STFT; Spectrogram; CNN; U-Net; t-SNE; and UMAP.

I. Introduction

Each music sample has unique context-dependent audio characteristics, making audio classification a challenging multi-modal task. Additionally, training classifiers requires annotated signals, which are difficult to obtain.

In this project, our aim is to analyse audio signals from different musical instruments. The extracted features are high-dimensional. We project them onto a two-dimensional plane to visualize their similarities and dissimilarities. For our experimental study, six musical instruments were selected, five of which are traditional Indian Instruments: Flute, Nadaswaram, Thavil, Santoor, and Veena. We also included the Western classical instrument piano and compared the audio characteristics of the music produced by the instruments. These traditional instruments possess unique tonal features. Flute and Nadaswaram are wind instruments that produce sound through air vibration, the flute being side-blown and made of bamboo, while Nadaswaram is a long wooden pipe with a conical end. Santoor (struck) and Veena (plucked) are string instruments, with Veena's dual resonators, add uniqueness to the music. Thavil, a South Indian percussion drum, has a hollow wooden shell with stretched leather membranes, and is played by hand or stick. The piano produces sound by hammering strings when keys are pressed.

Traditionally, Fourier Transform (FT) [1] and Fast Fourier Transform (FFT) [1] was used for signal analysis. But this approach cannot capture sequential contextual audio information. Advances in speech processing introduced techniques like Short-Time FT (STFT) [2] , Wavelet Transform (WT), and Mel-Frequency Cepstral Coefficients (MFCC) [3]. After using such audio feature extraction tools, machine learning techniques including deep neural networks that compress high-dimensional audio data. This effectively facilitated viewing music pieces, originiated from different instruments, as compact scatterplots on a plane. The overall plan is shown in Figure 1.

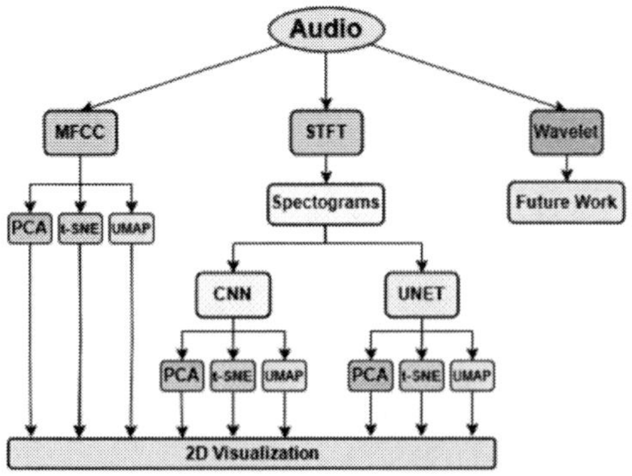

Figure 1. Overall plan for the Experiments.

MFCC features were extracted from the musical samples [3]. Using standard MFCC window lengths (25 milliseconds), even a few seconds of audio signal generate a high-dimensional feature vector. In this high dimension, the distribution of distances between samples exhibits low variance, making two-dimensional visualization ineffective.

We used a second step of feature compression using deep neural network. The effectiveness of the proposed methods was validated through several experiments by projecting the data onto two dimensions [4]. For spectral analysis, we used the

Short-Time Fourier Transform. We converted the STFT features into spectrogram images [2]. These spectrograms serve as visual representations of the music features. To extract features from spectrogram images, we used a CNN model [5], and a U-Net model [6]. CNN and UNet were trained on STFT spectrograms. Features were extracted from the output of filter layers of the CNN where they are input to the dense classification layer. Similarly, features were extracted from the bottleneck layer in UNet. Section III details how CNN and U-Net architectures extract features from images through their distinct approaches.

To visualize music signals on a two-dimensional plane, we used PCA (a linear method), t-SNE, and UMAP. MFCC features are directly fed into the above three visualization tools.

The remainder of this paper is organized as follows. Section II reviews related work. Section III describes the methodology, including data collection and pre-processing, feature extraction, and the proposed solution. Section IV presents the experimental results and their analysis. Finally, Section V concludes the paper and discusses future research directions.

II. Related work

The Previous works on the Visualization of audio sample characteristics are discussed below.

The authors used three different datasets in their work reported in 2024 [7]. Two datasets with 10 classes and an augmented version of one (using pitch shifting, time-stretching, and noise) were used. MFCC features were extracted, CNN and RNN-LSTM models were trained. CNN performed better on smaller datasets, while RNN-LSTM excelled on larger ones.

In the work reported in 2020 [4], the authors experimented with audio data of 10 classes, extracting MFCC features. They visualized these high-dimensional features using PCA, t-SNE, Iso-Map, and SOM. t-SNE produced well-separated clusters. SOM showed slight separation, while Iso-Map failed to capture meaningful structure. The conclusion was that Iso-Map failed to work with this high-dimensional data.

In another work on the audio classifier, reported in 2020 [5], the authors used a public dataset and converted the audio signals into Mel power spectrograms. They applied two approaches to capture features: a CNN model trained from scratch and a pre-trained VGG19 model using transfer learning. Both models performed well. The CNN model trained from scratch slightly outperformed the VGG19 model.

III. Proposed Methods and Experiments

This section outlines the paper's workflow, including data collection and preprocessing, feature extraction, projection of higher dimension into 2D, and the proposed method, as detailed below.

A. Data Collection and Pre-processing

Audio samples are collected from open public platforms like YouTube and recorded media, ensuring each sample captures the instrument's unique tonal and spectral characteristics without background noise or audio from other instruments.

We collected 180 audio samples, 30 samples per instrument, using YTMP3 and converted them to MP3. We processed them with Clideo, Clips were segmented into 30–45 seconds, then converted to WAV, ensuring high-quality data for analysis.

B. Feature Extraction

1) MFCC Feature Extraction: MFCC feature is a standard for audio analysis in music, speech recognition, and speaker identification. Pre-processing standardizes samples to 30 seconds by padding or trimming, followed by sampling at 44,100 Hz for high-quality signal preservation.

The MFCC extraction process begins with pre-emphasis to enhance high-frequency components. The 30-seconds audio is segmented into 25ms non-overlapping frames (1,201 frames, each with 1,103 samples). A Hanning window is applied to smooth edges and reduce distortions. The Discrete Fourier Transform (DFT) converts the signal to the frequency domain, capturing spectral characteristics. A Mel filter bank mimics human hearing by dividing the spectrum into 26 bands, reducing dimensionality while retaining essential information. A logarithmic transformation follows to compress the dynamic range. Finally, the Discrete Cosine Transform (DCT) decorrelates Mel-spectral coefficients, retaining the first 13 MFCCs used for classification.

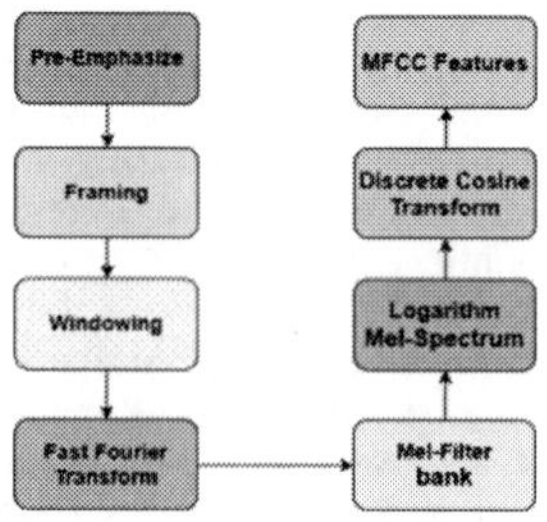

Figure 2. The process of MFCC Feature Extraction.

The MFCC extraction process is shown in Figure 2. Each 30-seconds sample is converted into 13 MFCCs × 1,201 frames and flattened into a 1-D vector of 15,613 elements. MFCCs capture essential audio characteristics, preserving tonal, timbral, and rhythmic features for classification and visualization.

The dataset for each musical instrument consists of 30 samples, each with 15,613 values, resulting in a data matrix of 30x15,613 for each instrument.

2) STFT Spectrogram Generation: The audio waveform of 30 seconds is divided into equal parts with a window of size 25 milliseconds. Each segment contains 1,102 samples. DFT of each segment is computed using the Fast Fourier Transform (FFT). The result of the FFT for each segment represents the frequency content of the audio within that window. These frequency domain representations are then concatenated to form the spectrogram image. The spectrogram displays the frequency spectrum where the intensity of a frequency is converted into brightness. The images corresponding to sequential windows provide a view of the audio spectral characteristics over the duration of the signal [8]. Figure 3 shows the STFT spectrograms of each musical instrument capturing the tonal and spectral characteristics of the instruments as images.

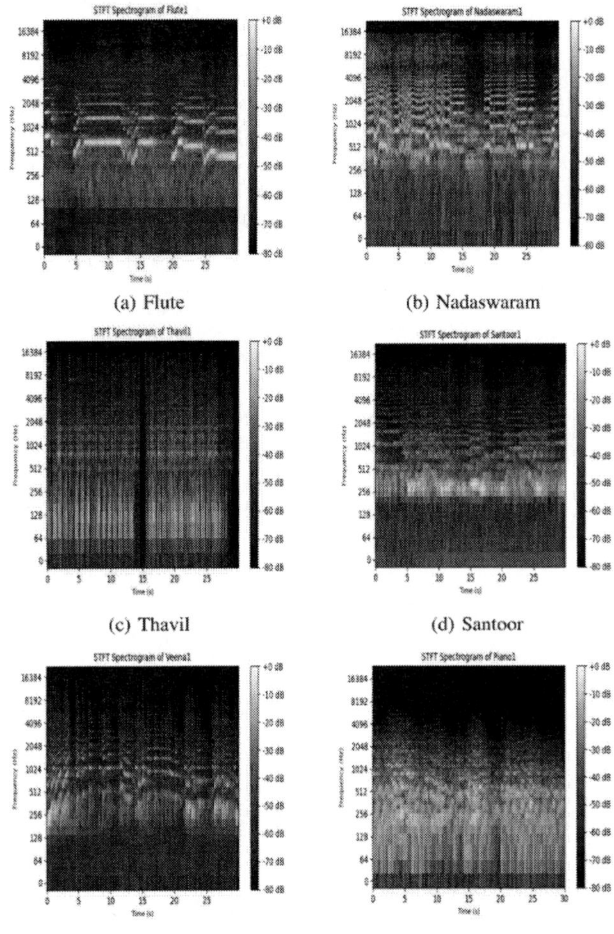

(a) Flute (b) Nadaswaram (c) Thavil (d) Santoor (e) Veena (f) Piano

Figure 3. Sample Spectrograms for each musical instrument.

C. Projection of higher dimension into 2D

1) Principal Component Analysis (PCA): PCA is a lightweight linear dimensionality reduction algorithm that identifies variance directions (eigenvectors) from the covariance matrix, which is symmetric with orthogonal eigenvectors. Projecting data onto the first two eigenvectors in 2-D highlights key data distributions [9]. In our project, PCA's effectiveness depends on the compression algorithm applied to MFCC or STFT data. We noted that UNet-compressed STFT spectrograms form well-clustered visuals even with PCA.

2) t-Distributed Stochastic Neighbor Embedding (t-SNE): t-SNE [10] is a non-linear dimensionality reduction technique, preserves the local structure of high-dimensional data by converting distances into probabilities and placing similar points close in lower dimensions. When applied to high dimensional feature space, t-SNE effectively captures complex patterns, revealing distinct clusters of audio data from musical instruments and uncovering structures.

3) Uniform Manifold Approximation and projection (UMAP): UMAP [2] is a non-linear dimensionality reduction technique that preserves both local and global structures, making it more effective for visualizing high-dimensional data in 2-D. By modeling data relationships as a graph and maintaining these connections in lower dimensions, UMAP faithfully presents complex distributions in low dimension.

D. Proposed Method

To visualize the audio features on a two-dimensional plane, we employed three visualization algorithms.

1) Visualization of MFCC Features: The MFCC features are extracted from the audio signals, resulting in a dataset with dimension 15,613 from every music piece of 30 seconds duration. In total, we have 180 samples for 6 instruments. Then we directly visualized them using PCA, t-SNE and UMAP.

2) Feature Extraction using CNN: The spectrograms of audio samples are used to train a CNN model with labels of the musical instruments. The CNN model architecture includes two convolutional layers, each followed by max-pooling layers, a flatten layer, and a couple of dense layers.

In Figure 4, the architecture of the CNN model used for training is illustrated. The input to the model is a spectrogram of size 400x600x3. The first convolution layer is with 16 filters to extract features, resulting in an output dimension 400x600x16. A MaxPooling layer with a 2x2 kernel reduces the spatial dimensions to 200x300x16. This is followed by a second Convolution layer with 32 filters and after applying 2X2 size max pooling, the resulting output is reduced to 100x150x32. The output is flattened into a vector and fed into a dense layer classifier. Since the musical instruments are known, the network is trained as a supervised classifier with feature vectors as input.

The extracted features are visualized using the visualization techniques: PCA, t-SNE, and UMAP. Feature vector scatter plots are projected on a 2-D plane to visualize the similarities and distances of the audio signals.

3) Feature Extraction using UNet: UNet which was proposed for medical image segmentation, is used to compress the image features. The UNet architecture consists of encoder and decoder structure: the encoding part uses convolutional layers followed by max-pooling layers to extract features and reduce dimensions, while the decoding part employs upsampling layers to reconstruct the input. Essential compressed audio features are available in the bottom layer of the UNet.

In Figure 5, the spectrogram of size 400×600×3 is input to the UNET model. Initially, two Convolution layers with 32 filters are used, followed by MaxPooling with a pool size of 4×4. Next, two more Conv2D layers with 64 filters are applied, followed by another MaxPooling operation. After that, two additional Convolution layers are applied one with 128 filters and the other with 64 filters leading to the bottleneck layer, which captures the encoded representation of the input.

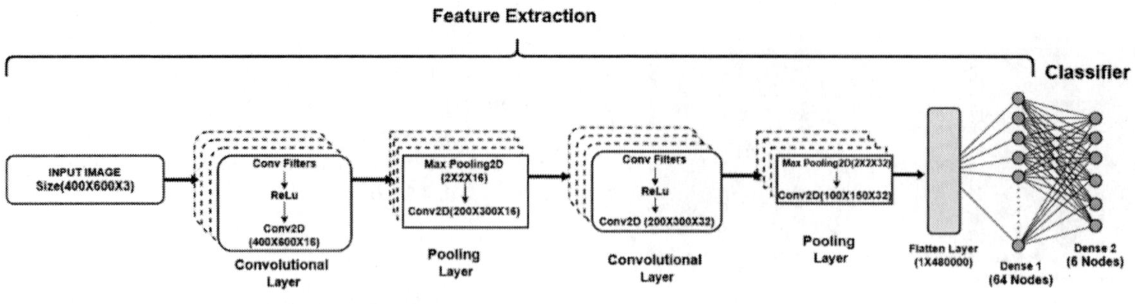

Figure 4. Architecture of CNN Classifier Model.

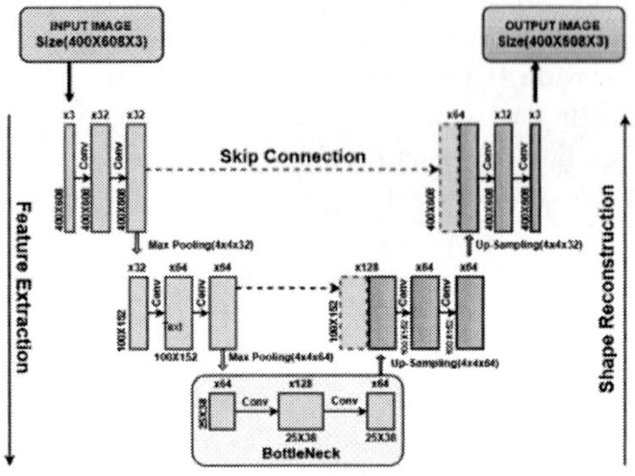

Figure 5. UNet Architecture.

From the bottleneck layer, the features are upsampled using a transposed convolution operation with 64 filters of size 4×4. These upsampled features are concatenated with the previous layer from the bottleneck. The process of Convolution, Upsampling, and concatenation is repeated for feature reconstruction. Finally, this process produces the reconstructed image. It is an unsupervised method.

The features from the bottleneck layer are extracted and used to visualize the data in a 2-D scatter plot using PCA, t-SNE and UMAP.

First, we train the UNet using data from all six classes. The compressed features from the UNet bottleneck, from different classes, are superimposed in the feature space and cannot be projected as separate clusters on 2-Dimensional plane. In the next experiments, we trained UNet separately with individual classes of features. After this training, the features from UNet bottleneck layer are used. The visualization algorithms projected them into clear isolated groups.

IV. Experiment and Results

In this section, we present the visualization results of MFCC features, and the extracted features from CNN and UNet.

A. *Visualization of MFCC Features*

To visualize high-dimensional MFCC features on a 2-dimensional plane we used PCA, t-SNE, and UMAP. The resulting scatter plot of 30 × 6 datapoints are shown in Figures 6, 7, and 8.

In Figure 6, the scatter plot for different instruments are completely mixed up. In Figure 7, we observe that Piano and Thavil samples are far apart and compact. It is as expected, because they produce very different types of sound. Veena samples are compact but mixed with Flute, Santoor, and Nadaswaram, which are overlapping and spread out.

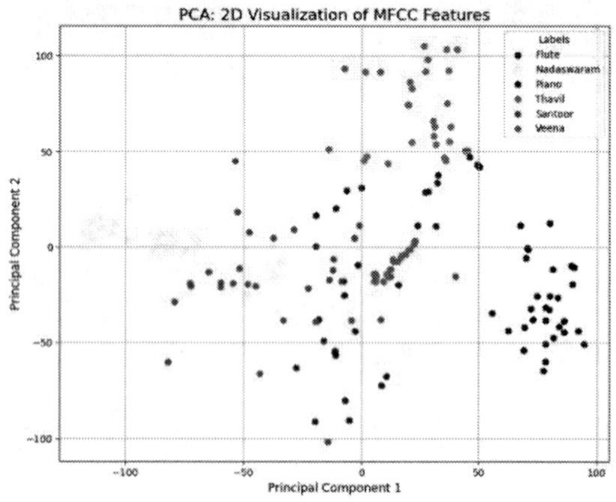

Figure 6. PCA visualization of MFCC features.

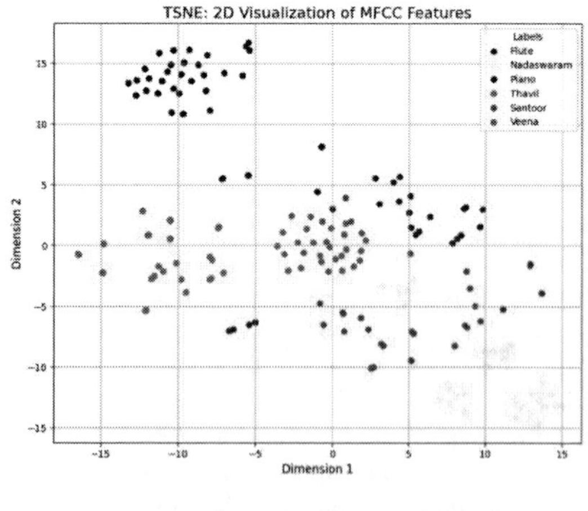

Figure 7. t-SNE visualization of MFCC features.

In Figure 8, the Piano samples are far from the others, while the rest form compact clusters. But these clusters are close to each other due to small interclass distances. The overall conclusion is that the MFCC feature, used directly as input to the visualization software, does not result in clear clusters.

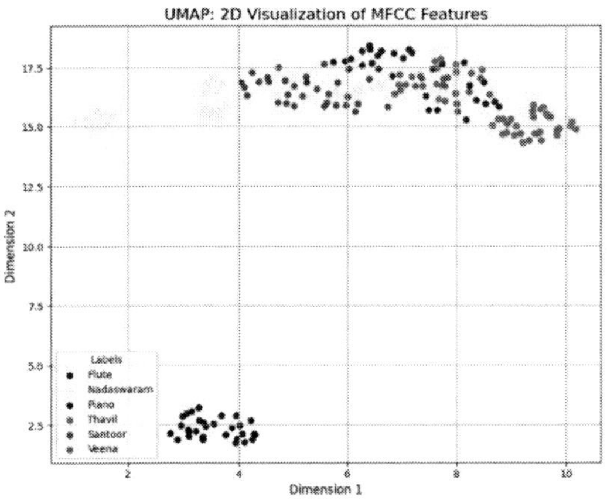

Figure 8. UMAP visualization of MFCC features.

B. Visualization of Extracted features from CNN model

Features extracted from the output of CNN, i.e., input to the dense layer for classification, are much lower in number than the MFCC features. These extracted features are fed into PCA, t-SNE, and UMAP for visualization. The results are shown in Figures 9, 10 and 11.

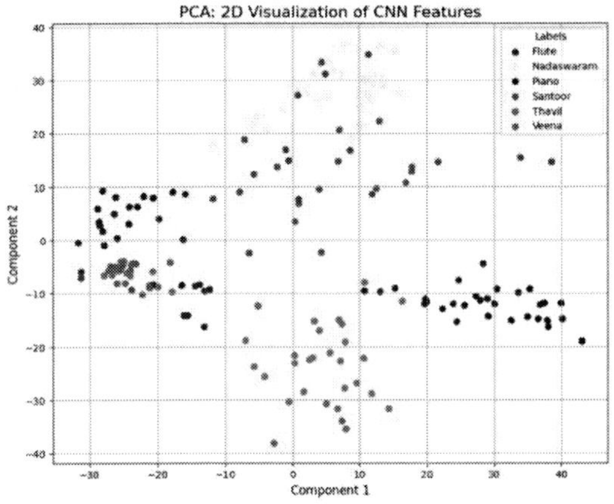

Figure 9. PCA visualization of CNN features.

In Figure 9, the thavil, nadaswaram, and piano samples form well-separated clusters though the sample points are scattered over a wide area. The santoor samples overlap with the nadaswaram cluster, suggesting some similarity in their features. The flute and veena clusters are positioned very close, indicating the related characteristics of the two instruments.

In Figure 10, all the music samples are clearly separated with large inter-cluster distances. The clusters are fairly compact, i.e., intra-cluster distances are not large. The piano and veena samples form compact clusters.

When UMAP was used for the 2D projection, as shown in Figure 11, we got compact non-overlapping clusters with clear large inter-cluster distances. The santoor and nadaswaram clusters are close to each other.

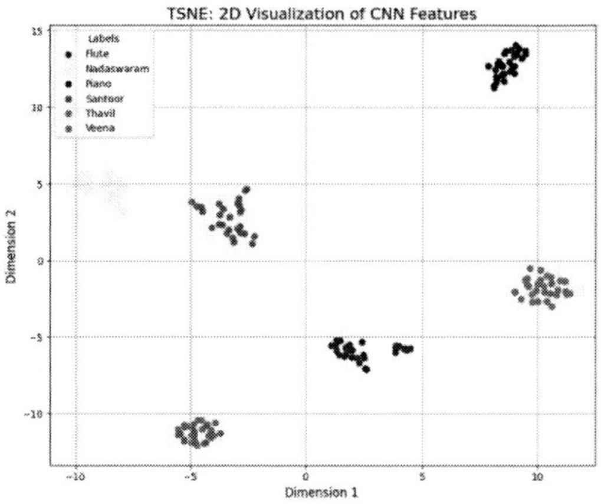

Figure 10. t-SNE visualization of CNN features.

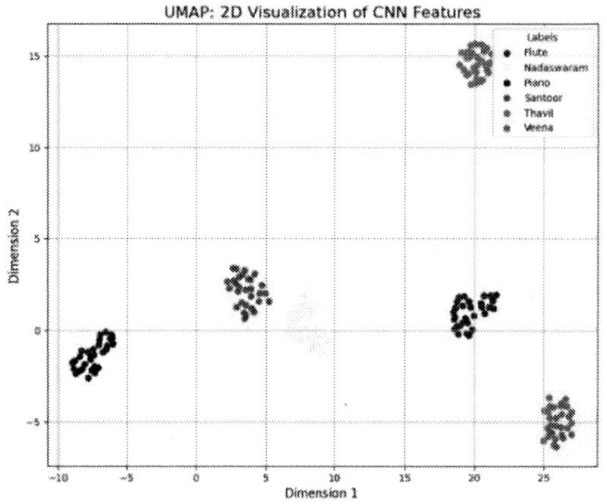

Figure 11. UMAP visualization of CNN features.

C. Visualization Results of UNet Features

The STFT spectrograms features were input into the UNet model and features at bottleneck layer were extracted. Thus, the original STFT features are compressed, and more abstraction is achieved at the UNet bottleneck. These compressed features are then used to visualize the data as scatter plot on a 2D plane using PCA, t-SNE and UMAP. The scatter plot results are shown in Figures 12, 13 and 14.

In Figure 12, we got clusters in which samples of every instrument are very compact. Veena and santoor clusters are very close to each other, which is quite contrary to what we got using UMAP and t-SNE. Thavil and piano cluster distances also close. Finally, nadaswaram and flute clusters are very far from the remaining clusters. Two things are to observe here, (1) the first eigenvalues are large and the second eigenvalues are around one-third of the first eigenvalues, (2) the interclass distances are very different from t-SNE or UMAP results, whereas the t-SNE and UMAP results are similar. This is due to linear projection with PCA.

In Figures 13 and 14, we got very well separated clusters where the intra-class distances are small resulting in compact

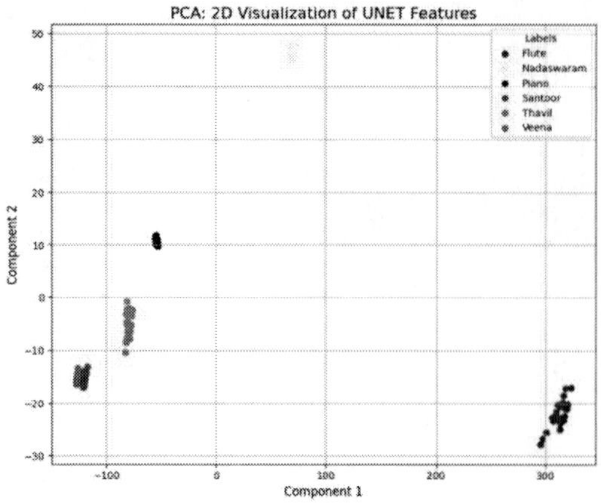

Figure 12. PCA visualization of UNET features.

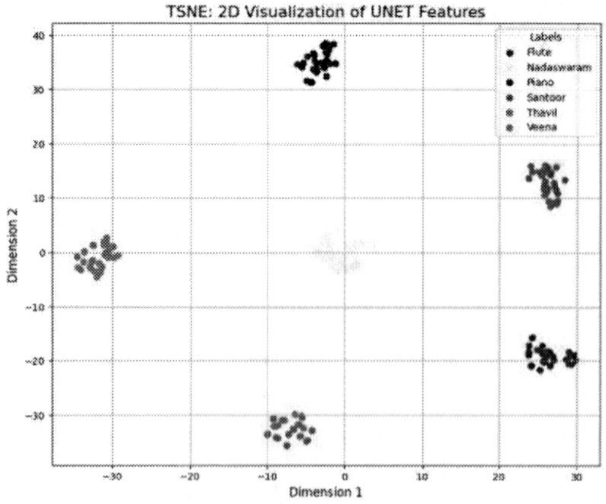

Figure 13. t-SNE visualization of UNET features.

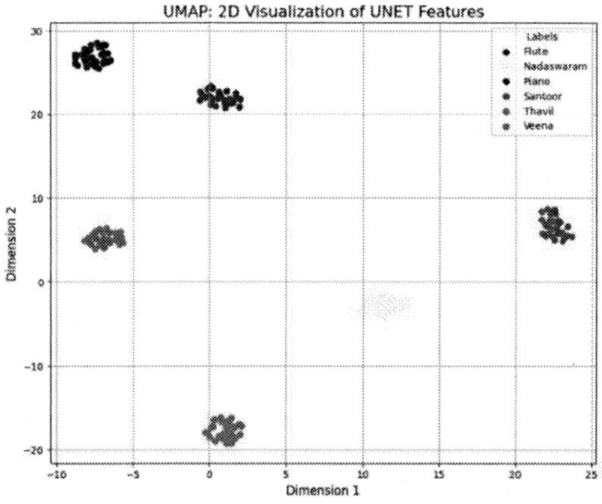

Figure 14. UMAP visualization of UNet features.

clusters. The inter-class distances are as expected and they are similar for the two visualization algorithms.

TABLE I
COMPARISON OF VISUALIZATION TECHNIQUES BASED ON SILHOUETTE SCORES

Visualization Technique \ Feature Selection	MFCC Features	STFT Spectrogram Features (CNN)	STFT Spectrogram Features (UNet)
PCA	32.85	35.21	76.50
t-SNE	37.37	83.99	74.60
UMAP	40.78	81.02	74.64

The Silhoutte score, which is the ratio of interclass and intraclass distances, are displayed in Table I.
PCA demonstrates moderate performance for MFCC Features and STFT spectrogram features using CNN but performs significantly better for the STFT features using UNET. t-SNE and UMAP outperform PCA for non-linear feature distributions, with t-SNE achieving the highest silhouette score for STFT Features using CNN and UMAP performing best for MFCC. Both t-SNE and UMAP show similar performance for the STFT features using UNET, indicating their suitability for high-dimensional feature visualization.

V. CONCLUSION AND FUTURE WORK

This study aims to find the correct tools to successfully visualize complex audio signals from musical instruments using machine learning and deep learning techniques. MFCC and STFT features were extracted and used to visualize their scatterplots on a two-dimensional plane by PCA, t-SNE, and UMAP. STFT features were converted to spectrograms, and Deep learning models CNN and UNet, were used to obtain a compressed version of the spectrogram image features. To visualize them in 2D, t-SNE and UMAP gave the best results, showing well-separated clusters.

It is difficult to quantify the correctness of the results. For further investigation, we will

- Find the first few eigenvalues to check how fast the eigenvalues are diminishing and how that is reflected when the data is projected on the plane of the first two eigenvectors.
- Compare the interclass distances resulting from three different visualization algorithms, and whether the relative distances from different methods are similar or not.
- Implement wavelet transform to extract music features, and then use wavelet spectrogram like, STFT spectrogram, and compare the results.
- Implement SOM as a tool for 2D visualization.

We will also extend this work for music generation combining music generated by different instruments.

REFERENCES

[1] M. Müller, "The fourier transform in a nutshell", in Aug. 2015, pp. 39–57, ISBN: 978-3-319-21944-8.
[2] L. McInnes, J. Healy, and J. Melville, "Umap: Uniform manifold approximation and projection for dimension reduction", *arXiv preprint arXiv:1802.03426*, 2018.

[3] S. M. M. A. Hossan and M. A. Gregory, "A novel approach for mfcc feature extraction", *2010 4th International Conference on Signal Processing and Communication Systems*, pp. 1–5, 2010.

[4] T. Pál and D. T. Várkonyi, "Comparison of dimensionality reduction techniques on audio signals.", in *ITAT*, 2020, pp. 161–168.

[5] B. Zhang, J. Leitner, and S. Thornton, "Audio recognition using mel spectrograms and convolution neural networks", *Noiselab University of California: San Diego, CA, USA*, 2019.

[6] O. Ronneberger, P. Fischer, and T. Brox, "U-net: Convolutional networks for biomedical image segmentation", in *Medical image computing and computer-assisted intervention–MICCAI 2015: 18th international conference, Munich, Germany, October 5-9, 2015, proceedings, part III 18*, Springer, 2015, pp. 234–241.

[7] K. M. Rezaul *et al.*, "Enhancing audio classification through mfcc feature extraction and data augmentation with cnn and rnn models", *International Journal of Advanced Computer Science and Applications*, vol. 15, no. 7, pp. 37–53, 2024.

[8] E. Wesfreid, "Preprint september 18, 2013 stft time-frequency visualization application to sound signals",

[9] S. Battaglino and E. Koyuncu, "A generalization of principal component analysis", in *ICASSP 2020-2020 IEEE International Conference on Acoustics, Speech and Signal Processing (ICASSP)*, IEEE, 2020, pp. 3607–3611.

[10] L. Van der Maaten and G. Hinton, "Visualizing data using t-sne.", *Journal of machine learning research*, vol. 9, no. 11, 2008.

Privacy-preserving Data Sharing Collaborations: Architectural Solutions and Trade-off Analysis.

Michiel Willocx, Vincent Reniers, Dimitri Van Landuyt,
Bert Lagaisse, Wouter Joosen, Vincent Naessens
DistriNet - KU Leuven
Gent and Leuven, Belgium
firstname.lastname@kuleuven.be

***Abstract*—Businesses and governments possess vast data with potential for analytical insights in areas like business intelligence (consumer behavior, business solvability) and governmental insights on population (crime, fraud). However, two primary challenges hinder the adoption of data-driven analytics: the lack of in-house expertise and the absence of sufficient data, which often requires collaboration with third parties. Such partnerships, especially involving Machine Learning (ML), raise concerns due to the sensitive nature of the data. This paper outlines two realistic use cases and proposes two privacy-preserving data sharing architectures tailored for business-to-business and government-to-business contexts. The first architecture uses de-identification techniques before and during data transmission, while the second assumes an already existing baseline ML model to test and refine predictions without sharing data. We present an in-depth analysis and evaluation of these architectures focusing on their complexity, trust requirements, and data-sharing efficacy.**

Keywords-privacy enhancing technologies, data collaborations, anonymity, utility

I. Introduction

Nowadays, companies and organizations have widely adopted the practice of collecting massive amounts of data during daily business activities. Businesses increasingly recognize the value of this data, or as commonly said "data is the new gold". Companies apply data for targeted marketing campaigns, to optimize the manufacturing process and inner workings, or even to increase customer satisfaction. Not only businesses, but also governments are increasingly interested in looking into ways to further collect or apply existing data. Governments are already sitting on vast amounts of data that can be put to work, for example to detect social fraud. For example, the case in which people benefit from social housing while in fact already owning (foreign) housing property [1]. Similarly, governmental data can be used to further improve crime fighting effectiveness (e.g., identifying problematic areas) [2] and for predictive analyses [3]. In order for governments and organizations to be able to perform these analyses, they first require (i) sufficient data, and secondly (ii) the important know-how to process this data. However, many business and governmental bodies are often lacking in both areas. While many organizations already possess a significant amount of data, more external data is often required in order to build qualitative prediction models. This data is to be acquired from third parties. Moreover, building these models (ML or otherwise) is often no easy feat, requiring expertise and experience to establish such models, and to subsequently evaluate and validate their accuracy. Many businesses therefore rely on third parties (i.e., data analytic parties) specialized in performing data analytics and building ML models for this.

In practice, organizations engage or desire to participate in a data sharing ecosystem that is highly beneficial to all parties. Such a data ecosystem involves close cooperation between different data owners on one hand, and between data owners and the ML party on the other hand. The benefits of sharing data are typically win-win situations, although establishing these data sharing collaborations comes with key problems related to both privacy and trust. The privacy problems are related to the fact that the data shared between parties often contains sensitive and/or personal information. The GDPR regulation [4] states that the type of personal data can only be shared when sufficiently anonymized. This means that records — or in some cases even attributes — in the shared dataset may no longer be linkable to an individual. In most cases, these types of issues can be tackled with state-of-the-art privacy models such as k-anonymity [5] and l-diversity [6]. Furthermore, the information may also be sensitive to the company, involving details on their inner workings (e.g., customer base), results (e.g., sales), or on collaboration with other companies, and sharing these data may impact their competitive advantage or reputation. Companies are very reluctant to share this type of information with their direct competitors, requiring significant trust, even if the result of the analysis of the data over all parties involved could be beneficial for all parties. Similarly, governments also posses data which may be highly beneficial when analyzed further by third-party analytical parties, yet this may not negatively affect the trust citizens have in their government and the safekeeping of their data.

In cases where a data provider is sharing data with a data analytics party, it sometimes suffices that both parties sign a non disclosure agreement. Even in this case, reducing the amount of required trust in such collaborations is desirable. New data collaboration strategies are required to solve these situations, where the end goal is the sharing of data for the purposes of gaining analytical insights via ML. In particular, such governmental-business or B2B data sharing require in-depth analysis of the requirements involved, as well as the technological solutions present to limit issues related to privacy and trust.

In this paper, we present two real-life use-cases from our collaboration with industry partners. In these cases, data

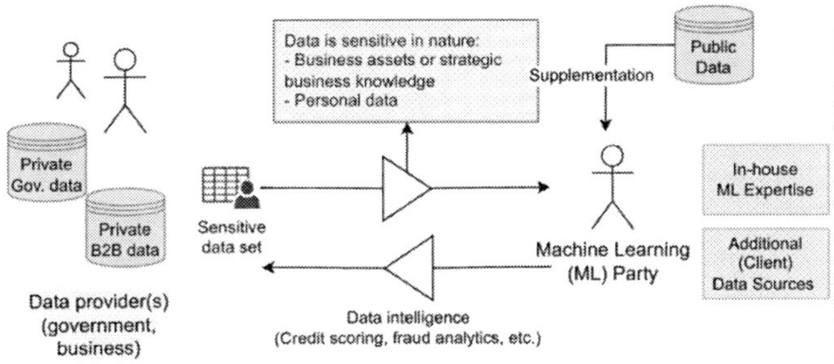

Figure 1. Use cases overview: Single or multiple data provider(s).

collaborations between different parties (data owner(s) and data analytics party) are required to solve important societal and economical problems. The specific nuances of each use case are laid out in detail. Next, two reference architectures are presented that allow to perform this type of data collaborations. The first strategy focuses on purely statistical methods, while the second one combines cryptographic constructs and statistics. The main advantages and disadvantages of each approach are evaluated and discussed in-depth.

The remainder of this paper is structured as follows. Section II introduces our motivating industry use cases and their requirements. Subsequently, Section III proposes architectural solutions, and the respective technological solutions that can be applied for each architecture. Section IV provides an in-depth comparison of both architectures via a trade-off analysis on aspects such as approach, complexity and implementation effort, and trust assumptions. Finally, Section V positions our work in the state-of-the-art, and Section VI concludes.

II. Motivating industry use cases

The research presented in this paper is driven by the use cases of two industry partners involving B2B and G2B data sharing for the purpose of ML analytics. In each use case, data is provided to the *machine learning (ML) party* by either one or *multiple data providers* (e.g., companies, organizations or governmental institutions), as shown in Figure 1. The ML party processes this data to generate insights via ML model creation and training, which they subsequently offer as products to the *ML party's customers*. The incentive for the *data provider(s)* is either of monetary value, or to acquire improved insights using their own data. Typically, the data provider lacks either (i) know-how to generate such analytical insights in-house or (ii) lack additional data gathered from a multitude of sources to do so. For example, it's possible that the ML party may incorporate additional collected data from private or public sources to enrich the data provider's data, and which enable it to in fact generate these insights. In both of our cases, the data provider is in fact also the end customer, which is the direct beneficiary of the obtained insights.

A. Use cases

We start off first by explaining the common denominator between all use cases, regarding their willingness to share data, which has to be abetted with privacy-preserving data sharing techniques. Subsequently, we detail our uses cases.

1) Problem statement: Willingness to share privacy-sensitive data: The common denominator between both use cases is that the data providers wish to share data, but are inhibited by the sensitive nature of the data. The data to be shared may involve information about other companies, internal company information that when leaked may for example impact the company's competitive standing or reputation. The data may also come from governmental sources, or involve information on data subjects that should not be disclosed. Therefore, typically a relationship of trust has to be established between the data provider and the machine learning party before data is shared, for example via contractual agreements. In this research, we want to avoid or minimize the degree of trust required before sharing data, by either applying privacy tactics and abstracting the data to a degree, or keeping the data on-site and steering the ML party's generated insights. Such tactics can enable B2B data sharing without a significant vote of trust in the ML party, by managing what is shared, and what can be learnt from it. Finally, the customers of the end ML party's trained model, should never be able to infer on which data was used to train the data set. In the next subsection, we more precisely formulate the nuances of each use case, after which we will translate these use cases into concrete requirements for both data provider and the ML party.

This paper identifies two theoretical dimensions that are possible in a data sharing ecosystem, namely regarding (i) sensitive nature of data (e.g., personal or non-personal data), and secondly (ii) number of data providers. The first dimension involves the nature of the sensitive data that is being shared, which may either be personal or non-personal data, and which influences the solution architecture on the potential requirement for GDPR-compliancy. The second dimension involves the number of data providers which are involved for one case, which may either be one entity providing all data required, or many entities providing pieces of the puzzle. The latter dimension may also impact potential solutions, for example when there is only one data provider the source is already self-evident. In the case of multiple data providers, possible solutions may rely on techniques such as e.g., multi-party computation to enable data intelligence gathering. These dimensions theoretically cover numerous use cases involved with data sharing, which amount to several possible combinations, for example non-GPDR data and many-to-ML data providers. Our two industry use cases that motivate this work represent different characteristics in these both dimensions. These use cases may practically apply to a wide range of B2B or G2B data sharing scenarios. We will detail the nuances of each of these two specific use cases, their dimensions, and analyze the requirements of each stakeholder, more specifically the requirements of the ML party or data provider.

2) Industry use case A: Single governmental data provider and ML party (G2ML): In our first use case A, which motivates this work, the data provider is a single entity, namely the government, that contains all required data on which the ML party can generate insights. Governments can have vast pools of even public data that can be accessed, such as information on the population, maps (building zones, agriculture), environment (e.g., pollution). There may also be more restrictive information involved, for example stored in police or judicial databases. Based on such vast pools of both public and private data the government can generate a multitude of insights, to improve their enactment on for example environmental polluters, or to identify problematic areas in society. Governments typically lack in-house expertise to generate such insights and therefore ideally rely on external an ML party. In this case, when private data is involved, the government has to either establish a trust relationship with the ML party, although this can be tricky. Therefore, we aim to provide technological solutions such as privacy-preserving techniques and architectures to limit such trust requirements in the ML party. Certain other requirements may also apply, such as requiring the ML party not to disclose any of the shared data, or even when privacy techniques are applied, to not disclose the learned ML insights.

3) Industry use case B: Joining business data to generate sector insights (B2ML): In our use case B, multiple industry companies are data providers to a single machine learning party. The data shared in the many-to-one relationship can be of non-personal sensitive data, or data which has to be GDPR compliant. For example, supermarkets may store a lot of data on the shopping behavior of customers, such as products frequently bought together, or even have the potential to store very specific information on a per-customer basis. These companies could contract a ML party to process their data and generate sector-wide insights regarding purchasing behavior, which could for example optimize advertising campaigns. In certain cases, companies are however not inclined to share the full details of the data set, or wish to omit certain person-specific aspects, and this then requires technological solutions to enable the B2ML data sharing process. Another requirement is that these companies may not wish to share data among each other, as this may impact their individual competitive standing. In addition, the ML party must be trusted to not disclose individual datasets to other competitors, or we can rely on privacy-preserving tactics to facilitate this requirement. In the next section, we generalize the requirements for each stakeholder, for example regarding the data provider and ML party. These requirements will guide our architectural solutions for privacy-preserving B2B or G2B data sharing.

B. Requirements analysis

We enumerate the requirements for each party of primarily the data provider, and the machine learning party that processes this data. These requirements relate either to privacy aspects, functionality, or non-functional requirements.

1) Data provider requirements (R_{DP}): The data providers are companies or governmental institutions, which feature a certain degree of willingness to share data, or are highly reluctant to do so unless certain privacy guarantees are met. This property of willingess to share data will impose more subtle or stringent privacy tactics. For example, the data that is shared may not be leaked to any other party than the ML, or even more stringent, the ML party itself may not be able to deduct which subjects were in the shared data set, a property referred to as unidentifiability or unlinkability. These privacy tactics ideally won't impede certain functional requirements the data provider desires in return for providing the data. For example, the data shared must yield the data provider itself with additional insights learned by and from the ML party.

2) ML party requirements (R_{ML}): In terms of functional requirements, similarly the ML party will want to acquire as much data as possible, or sufficient data that enables them to generate ML insights. These insights can be of use to their customers, which in our use cases is the data provider itself. In terms of privacy requirements, which may be imposed by the data provider, the shared insights cannot be used to deduce which data subject was involved in the training set (membership inference). In addition, the ML party wants to establish a certain degree of trust in their process, which can come from the privacy tactics that are applied before sharing the data with the ML party. Such trust in algorithms, rather than the parties themselves, will promote future data sharing collaborations. Finally, regarding the insights generated, the process of the ML model may be subject to intellectural property rights (IPR), and therefore details regarding the ML model are ideally not disclosed to any of the other stakeholders, as this may compromise their core business. Optionally, the insights when disclosed are also of a sufficient quality when shared (i.e., good accuracy), to promote the reputation of the ML party.

3) ML party customer requirements (R_C): The customer of the ML party is typically the data provider itself, but also other organizations or institutions that can benefit from these insights, or generate additional insights using additional combinations of their data. The requirements for this customer is that the insights are sufficiently accurate, and potentially that these insights do not come with stringent privacy measures (for example imposed by legal laws). The latter could be the case when certain data subjects can be identified from the analysis, such linkability/identifiability is ideally not possible and a previously listed requirement of the data provider, imposed on the ML party.

4) Data subject requirements (R_{DS}): In our use cases, the data may or may not be subject to the GDPR legal framework as in most cases the data is related to organizations or institutions, and not individuals. An edge-case in this regard is the situation where a company is a one-man company, in which case data related to that organization is considered personal data [7]. In cases where personal data in involved, sufficient data anonymization should be applied in order avoid re-identification.

III. Architectural designs for privacy-preserving data sharing

We present two architectural solutions to meet with the general use case requirements outlined in the previous section. The driving factor to choose between both architectures is mainly the degree of willingness to share data by the data provider, and the individual architectural properties. Section III-A presents an architecture in which privacy-preserving tactics can be applied at the data provider-side before it reaches the ML party. Alternatively when dealing with many data providers, these privacy-preserving tactics can be applied once more and intermittently by a mediator. Section III-B details an architecture in which no large data sets are shared between data provider and the ML party. Instead, the ML party tests its predictions at the data provider, which only provides data in the form of minimal feedback to correct the ML model. We will detail each architectural design in-detail, followed by a discussion of their properties and respective trade-offs.

A. Architecture 1 – Sharing privacy-enhanced data

The first architecture to accommodate data sharing between data provider(s) and the ML party relies on statistical anonymization strategies. In this approach, the data is sent from the data provider(s) to the ML party. However, before transferring the data, it is anonymized by the data provider.

1) Core approach: Figure 2.A presents the situation in which one data provider shares data with the ML party. A client-side module (which can be provided by the ML party) locally performs de-identification operations on the data. These operations can include data scrubbing (deleting vulnerable records/attributes), pseudonymization [8] (replacing identifying attributes with a pseudonym), generalization (applying privacy metrics such as k-anonymity [5]) and noise addition. The required operations strongly depend on both the nature of data and the intended use case. Hence, for each case specific settings are required in the client-side privacy module. These settings require considerable insights in the data and the attacker model, and are therefore ideally created in collaboration between the data provider and a privacy expert party (e.g., which may be coincidentally be the ML party).

The major limitation of this approach is that it is only applicable with suitable datasets. Data collected by companies is often sparse or incomplete. In order to enable meaningful data de-identification it is desirable to first complete and enrich the data using external data sources. This data enrichment step in Figure 2.A(1.b) is a second task of the client-side module. Examples of data enrichment are replacing a social security number of individuals by a range of quasi-identifying attributes (e.g., date of birth, residence location, gender) or replacing the VAT-number of a company by relevant information (e.g., location, amount of employees, profit margins). The required data can be collected from either public and private sources, or provided to the client-side module by the ML party.

The first approach presented here allows the data owner to remove personal identifying information (and hence to share personal data in a privacy-friendly and GDPR compliant manner). It can also ensure that certain sensitive aspects of the data are also generalized away, although the input and output of sensitive data should be carefully curated by the data owner to ensure confidentiality. The main downside of this configuration is that the machine learning party can directly link the data to the data provider as it is originating from this source. Our next alternative on this architecture in Figure 2.B introduces a mediator to also further generalize between multiple data providers, but also to hide which data originates from which party.

2) Advanced approach with mediator: There are two major disadvantages of the core approach, the first is that data originating from the data provider is directly linkable to the data source origin. In addition, when multiple data providers are present, further data generalization and privacy-preserving tactics can be applied on the combination of these data sets.

These issues can be solved by introducing mediator in the system, such as a TTP (trusted third party). Figure 2.B presents the approach that includes a mediator, and as example we will assume the use of a TTP. In this setup, all data providers send the data to the TTP. This data can be provided to the TTP in a (partly) privacy-enhanced (e.g., de-identified) state. The TTP collects the data from the different sources, after which the records from the different data providers are merged and mixed. Next, the TTP performs a de-identification step on the data to ensure that the confidentiality of the data and/or the privacy of the data subjects are preserved.

The addition of a TTP allows a decoupling of the data from the data owner. In addition, it can provide an additional de-identification step, and when this is performed correctly and sufficiently, the ML party should be unable to link a record back to the correct data owner. However, this approach is based on the important assumption that the TTP functions correctly, behaves honestly, and does not collude with the ML party. Collusions with the ML party could allow the ML party to link records to one specific business, as would be the case for the core approach without mediator. The trust required by the data provider shifts from the ML party to the TTP. As the data still leaves the premise of the data provider, he might still be reluctant to allow this. A high degree of auditability could offer a solution in this regard. In this setup, the TTP could also be responsible for the data enrichment step (as described in the core approach). This is often even more desirable for the ML party, as the data used for enrichment (which can be intellectual property of the ML party) no longer needs be shared with the data provider.

In the architecture we just presented, proposed a method to share data after applying privacy-tactics at a client-side privacy module, and possibly additionally again at a privacy module located in an intermediate mediator. In these approaches, data is effectively still charged, and even after the deliberate steps and transformations to preserve privacy, this may still be undesirable by parties that are highly reluctant to share data. In our next architecture, we propose a method to improve the ML party's models without sharing data.

B. Architecture 2 – ML feedback-loop validation interface

The second architecture relies on cryptography in combination with statistics. The main advantage of this approach is that the data provider is no longer required to share any data with the ML party. This approach assumes that the ML party

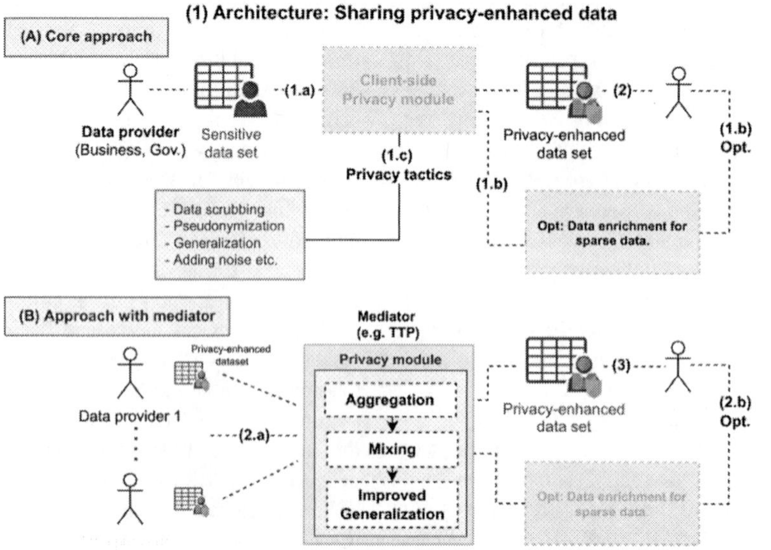

Figure 2. Architecture 1: Privacy-enhancing tactics on (non- or personal) sensitive data.

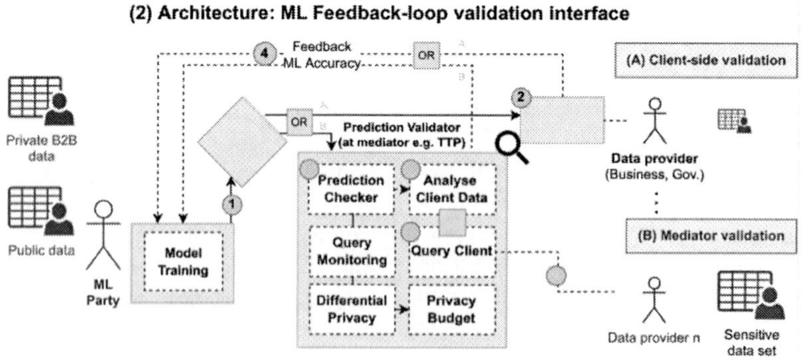

Figure 3. Architecture 2: Feedback-loop for validating or improving ML accuracy without sharing data.

is able to create a preliminary ML model. This model can be created based on a small amount of data retrieved from a data provider (under NDA or by applying Architecture 1) or by using publicly available data.

1) Core approach with the prediction validator module at the data provider: In this approach, the ML party provides a software module to the data provider(s), named the *prediction validator*. This module runs on the premise of the data provider and has access to the sensitive data of the data provider. The data itself never leaves the premise of the data provider. Next, the ML party makes predictions using its preliminary basic ML model, and sends these predictions to the prediction validator module at the data provider. Here, the validator compares the ML party's predictions to the sensitive data records. Based on this analysis, the validator is able to generate the feedback the ML party requires to enhance the ML model.

The kind of feedback and the level of granularity of the feedback depend on the type and sensitivity of the data involved. It is of major importance that the feedback does not leak the sensitive data of the data provider. Therefore it is not possible for the validator to provide the ML party with feedback about individual predictions. However, it is for example possible to give a general accuracy score about groups of predictions. For example, the ML party can group its predictions in confidence intervals, which allows the mediator module to give feedback about a group of predictions. Furthermore, this approach could also support typical machine learning metrics such as recall, precision and F1-score [9].

It is important to be aware not to disclose any sensitive information of the actual data set. By executing consecutive queries, the ML party could attempt to learn sensitive information. For example, if in two consecutive queries, a set of predictions is validated, but one query leaves out the prediction of one record, the ML party could easily learn the details of that one record. Query monitoring [10] prevents this type of unwanted behaviour. Moreover, by applying differential privacy [11], [12] to the output of the validator (adding noise), an additional layer of protection against data leakage is introduced, protecting against information leakage even if the ML party has (partial) knowledge about the sensitive dataset.

The presented approach requires the ML party to share its predictions with the data provider in order to retrieve feedback on these predictions. In an early phase of the training, these predictions may still be very immature and may lead to reputation damage for the ML party as the predictions may be potentially very inaccurate. Furthermore, these intermittent predictions may reveal part of the inner workings of the ML party's (protected) model, which may be subject to IP rights. This may undermine the ML its competitiveness if the data provider is able to steal (part of) the ML party's model. As a solution, we suggest an alternative approach with a validator module located at a mediator to avoid the need for the ML party to directly share its predictions with the data provider.

2) Approach with the prediction validator module in a trusted mediator: One of the main disadvantages of the core approach is that the ML party leaks its (preliminary) predictions to the data provider. This is especially important as both the ML parties in our industry cases listed this as a major concern. In order to avoid data leakage from the ML party to the data provider, the extended version of this approach moves the prediction validator module to a trusted mediator. Hence, the predictions of the ML party are no longer sent to the data provider. The mediator's main task is to compare the predictions of the ML party to the (sensitive) data of the data provider. In order to prevent the mediator from gaining access to the sensitive parts of the data, the ML party and the data provider can apply symmetric encryption to certain attributes before transmission. To achieve this, the data provider and the ML party must first exchange a shared secret key, of which the trusted third party has no knowledge. Comparisons can be made on the encrypted data without the need of additional technologies However, homomorphic encryption – a technique enabling (basic) operations on encrypted data [13] – can also support more fine-grained analyses of the predictions. By applying this approach, the trust required in the mediator is significantly reduced compared to Architecture 1. The only assumption that needs to be made is that the mediator operates honestly. Honest means that it does not tamper with the analysis results for the ML party, and that it does not share the input retrieved from one data provider with another, or data from a data provider to the ML party or vice versa.

IV. Architectural trade-off analysis

In this section we provide a comparison via a trade-off analysis on the properties and merits of both architectures in terms of approach, complexity (e.g., implementation effort), but also requirements on trust between all stakeholders. Specifically, we also detail which technologies, and for example privacy techniques, can be concretely applied towards the implementation of these proposed architectures.

A. Architectural comparison

Table I lists an extensive comparison between architectures 1 and 2, both when or when not using a mediator. This comparison is conducted in terms of the factual shared data, the required quality of shared data, the implementation effort for each architecture, and overall complexity. In addition, we detail the mediator's task in each architecture, as well as the overall followed approach, and eventual consequences for data linkability. The main goal of the proposed architectures includes that individual subjects cannot be identified from the data set. We discuss each aspect from Table I in turn.

1) General approach: In architecture 1, we privacy-enhance and actually share data with the ML party. In the second architecture, we do not directly share data, but instead provide feedback information (i.e., validation) on the ML party its trained model, namely on its prediction accuracy.

2) Optional mediator's involvement and task: In the first architecture, optionally, a mediator is involved to further generalize from multiple data providers', and consequentially hide the data source from the ML party. In the second architecture, the mediator is involved to either hide the data provider which validates the forwarded predictions by the mediator. Alternatively, the data provider can also entrust the data set to the mediator, which then assumes the responsibilities to provide feedback on ML predictions. The mediator has to however ensure that subsequent feedback responses do not reveal anything about the original data set, via techniques such as differential privacy and query monitoring.

3) Data linkability: In terms of data linkabilty, depending on the choice for the first architecture and the involvement of a mediator, either the ML party can directly attribute a certain data set to a certain provider, or the mediator is able to do this and hides such information from the ML party. In the second architecture, predictions cannot be linked to a concrete data set, only the feedback to a certain mediator or data provider.

4) Shared data and quality: The main distinction between both architectures is the willingness to share data, and architecture 1 is ideally suited for sensitive data which can be privacy-enhanced and still shared, whereas in architecture 2 only feedback is given on the ML model its accuracy. In the first case, we therefore need sufficient quantity and diversity of data as to enable the application of such privacy tactics.

5) Implementation effort and complexity: In terms of implementation, architecture 1 can make use of readily-available software libraries featuring privacy tactics, such as the ARX library [14]. The only difficult aspect is that the chosen tactics have to be carefully considered regarding their suitability on the involved data set, and their respective impact on privacy threats and remaining data utility. In contrast, the second architecture is more visionary, and requires careful consideration on how to provide feedback or validation of the ML's predictions, of which optionally this feedback can steer the training in a positive manner. In addition, this feedback should not reveal anything about the data provider's data set, which may require differential privacy and query monitoring, which consider previously released queries' and their respective feedback.

TABLE I. CHARACTERISTICS AND PROPERTIES OF THE ARCHITECTURES WITH- OR WITHOUT MEDIATION.

	Architecture 1		**Architecture 2**	
Shared data	Data sets which are privacy-enhanced (e.g., de-identification, generalization of attributes).		No datasets shared, only minimal feedback regarding the accuracy of the ML model.	
Data quality required	Needs sufficient data (e.g., minimum number of rows) to privacy-enhance data (e.g., generalization).		Doesn't need directly shared data. Feedback is given to validate the ML model, and potentially improve it.	
Implementation effort	Standard libraries available to apply the privacy tactics, such as the ARX library [14].		No ready-made available libraries/frameworks for such an approach.	
Complexity	Hand tailored selection of privacy tactics per use case.		Complex process to determine how to structure valuable and privacy-preserving feedback.	
Mediator's task	Generalization, de-identification, and other tactics of multiple already privacy-enhanced data sets.		Responding with feedback, querying or optionally collecting data from/to data providers.	
	No mediator (1.A)	**With mediator (1.B)**	**No mediator (2.A)**	**With mediator (2.B)**
Approach	Data is privacy-enhanced at the data provider and then sent to ML.	After privacy-enhancing at the data provider, additional generalization at mediator.	Data is not directly shared, but the prediction accuracy of the ML model is validated at the client side.	Data not shared directly, ML model validated at the mediator (entrusted the data or has to query client).
Data linkability	Data originates directly from the data provider, and is linkable to the source.	Removing direct link to origin of data per DP.	Predictions cannot be linked to a certain concrete data set.	Client can maintain its own data set, or share it with mediator.

B. *Trust model analysis*

Table II elicits the trust assumptions, which are generally minor, for all involved stakeholders, namely data provider, ML party, and optionally a mediator.

Data provider: Regarding the data provider, we assume that this provider acts honestly and shares a correct (privacy-enhanced) data, or in the case of the second architecture provides honest feedback on the basis of this data. In theory, this should be in the interest of the data provider himself, as he will typically rely on the insights gathered by the ML party, which is a win-win for both actors. In turn, the data provider could possibly attempt to reverse engineer the ML model based on the insights, although this could prove technically challenging, and is therefore a weak assumption. These insights are more valuable in the second architecture, which are presented intermediately, and the process of model learning could be more evident.

ML party: The trust assumptions placed in the ML party are more of a minor nature, as in the first architecture the data that arrives is already privacy-enhanced. Yet, potentially we could expect the ML party to not further disclose this privacy-enhanced data set. We expect the ML party to share honest insights gathered, but this could be facilitated or verified by the data provider on the basis of real world scenarios, or applicability in its own business processes. In the second architecture, no concrete data is shared, but insights which out of self-interest are ideally honest.

Mediator: As a mediator, many of the trust assumptions of the data provider and ML party are partially inherited. For example, we assume it forwards the original privacy-enhanced data set in architecture 1, or the corresponding feedback on the ML party its predictions when querying the client. Alternatively, when the mediator is entrusted the data set in architecture 2, we also assume it does not disclose this data (which may be a stringent requirement, although when sensitive we also do not expect the ML party to do this). In addition, in this case we assume a correct handling of the predictions. Furthermore, we also expect the mediator to hide the data source, more specifically the data providers involved.

1) Meta-data encryption: In our architecture, and when it is opted for a mediator, and specifically in the case of a trusted third party, we assume that this TTP is honest-but-curious. In Architecture 1, the data which arrives at the mediator is already privacy-enhanced, and is ideally further aggregated. The trust at this stage, is therefore mainly in the correct application of this method and the forwarding. In Architecture 2, the operations applied by the mediator are more complex, and he can have insight into the predictions passed by the ML party, as well as verifying these predictions by a query to the data provider (or when trusted against the data set provided to the mediator). In order to hide the insights that the mediator can gain into the process, both ML party and data provider can agree on a shared key to encrypt meta-data before sending it over the mediator. This will enable part of the task of responding to the prediction query by the data provider, or in reading part of the feedback by the ML party.

2) Trusted execution environments: Such encryption can not always be applied however, as the mediator may have to be actively involved in assessing whether the prediction is correct, and involved in the feedback process. Therefore such key values may have to be in readable format. In order to prevent the trusted mediator, which is assumed to be honest-but-curious, from gaining such insights, and to actually also relax this trust assumption we can integrate trusted executions

TABLE II. TRUST ASSUMPTIONS OF THE ARCHITECTURES WITH- OR WITHOUT MEDIATION.

	Architecture 1		Architecture 2	
Trust in	**No mediator (1.A)**	**With mediator (1.B)**	**No mediator (2.A)**	**With mediator (2.B)**
Data provider	Shares correct data set (win-win for insights), least possible noise. No reverse engineering of ML insights.		Provides only honest feedback. Keeps intermediate insights confidential.	Provides honest feedback and keeps intermediate insights confidential, or trusts true data set to mediator.
ML party	No data disclosure. Shares truthful insights.		Provides truthful insights out of self-interest.	
	With mediator (1.B)		**With mediator (2.B)**	
Trust required in mediator	Shares correct privacy-enhanced data, and only to ML party. Hides source of the data.		Mediator passes correct insights and feedback, or optionally keeps data confidential and provides feedback.	

environments. Trusted execution environments provide a secure tamper-resistant execution environment, isolated from – in this case – the rest of the mediator's own platform [15]. For example, Intel SGX [16] could be used to execute certain of the mediator's its functionalities. Subsequently, the data provider its data can be sent to this module encrypted, and will only be readable to the trusted execution environment.

C. Architectural selection process

The selection of one of the presented architectures for a specific use case is influenced by multiple factors, namely the quality and the nature of the data, the willingness of the data provider to share the data with the ML party and the willingness of the ML party to leak information about the ML model with the data provider. A first distinction is made between whether the data concerns sensitive company data, as this is a driving factor for the willingness of the data provider to share the data with the ML party. If the data is not sensitive, and does not contain personal data, no privacy enhancing tactics are required. If the dataset contains personal data that is not sensitive to the company, Architecture 1.A is advised when the data is suitable for anonymization (by default or after enrichment). Alternatively, when data anonymization techniques are not feasible, a variant of Architecture 2 is required. In the scenario where the data provider is reluctant to share the data with the ML party, Architecture 1.B can be applied if multiple data providers are available and the link between the data providers and their respective data can be severed by mixing (and anonymizing) data from multiple data providers. When this is not possible, a variant of Architecture 2 is advised depending on whether or not the ML party requires model protection.

V. RELATED WORK

Our work is situated within the domains of classical privacy-preserving techniques, data anonymization for ML purposes, and collaboration strategies in this context such as federated learning.

Data anonymization strategies: Many well-known data anonymization strategies are described and evaluated in literature. Privacy metrics such as k-anonymity [5] and its derivatives such as l-diversity [6] and t-closeness [17] are extensively studied, in the context of privacy [18], [19] as well as the theoretical [20] and the practical utility [21], [22]. Moreover, they are readily available in tools such as ARX [14]. Many real-life use cases have benefited from these types of metrics. For example, Jakob et. al. [23] described an anonymization pipeline to aid the gathering of medical data for research during COVID.

Anonymized data applied in ML: The applicability of anonymized data in machine learning applications specifically has also already been discussed by several papers. For example, Slijepcevic et. al. [24] and Carvalho et. al. [25] investigate the effect of applying metrics such as k-anonymity on the classification performance. Both works argue that it is hard to exactly predict the impact of privacy preserving operations on the accuracy of ML models, but find that the effects are manageable if the anonymization operations are not too harsh. The aforementioned data anonymization strategies are applied as part of the solution in Architecture 1, but require additional components in order to fulfill the trust requirements related to the sensitive nature of the data.

Protecting machine learning models: In the context of this paper, two important attack vectors on machine learning models should be considered. First of all, many papers [26], [27] have demonstrated that machine learning algorithms are often prone to leak data used in the training set. Two popular types of attacks are membership inference [28], [29] and attribute inference [30], [31]. Defenses against these types of attacks are proposed [32], [33] and should be considered in both architectures presented in this paper.

A second threat to ML models that is relevant in the context of this work is model stealing [34], [35]. As the machine learning model is intellectual property (and the core business incentive) of the ML party, the model should be protected against such theft. Several defenses have been proposed [36], [37], and should be implemented by the ML party.

Privacy-preserving querying: Within the context of privacy preserving data sharing, the concept of differential privacy [11], [12] is currently often presented as a one-size-fits-all solution. In contrary to the aforementioned data anonymization techniques, differential privacy is not a property of a dataset but of a function. Therefore, differential privacy is not directly applicable for the data sharing part of our industry use cases, as they require the actual records and not aggregates over multiple records. Differential privacy is however applicable in Architecture 2 in the mediator module, as it can prevent data leakage in the feedback to the ML party.

Alternative privacy preserving data collaboration strategies: In the realm of machine learning, federated learning strategies [38] are being proposed, allowing companies to collaborate towards a common machine learning model without the need to contribute their own data in one shared data pool. For example, Dayan et. al.[39] demonstrate the advantages of federated learning to create data collaborations in the context of a large COVID-19 clinical study across multiple countries and health institutions. Tools and frameworks such as Flower [40] and Sherpa.ai [41] support developers in implementing such strategies. However, it should also be noted that successful attacks have been performed on federated learning models before [42]. The business driver in our industry cases is the ML party, whose incentive is the financial benefit from commercializing the created machine learning models. Applying federated learning in our industry use cases would cut the ML party (and therefore the technology enabler) from the equation. Additionally, a federated learning approach would also rely on the data providers to set up such collaborations (and processing infrastructure) among themselves. Such solutions are therefore undesirable and unfeasible in our industry use cases. Another stream in privacy preserving data collaborations is found in the realm of cryptography, where techniques such as fully homomorphic encryption (FHE) [13] and secure multi party computation (SMPC) [43] have gained traction. FHE allows computations to be executed on encrypted data. In this work, FHE can be applied as one of the building blocks in Architecture 2 in order to support more complex model validation in the mediator module and to enhance the feedback towards the ML party. Note that the set of available operations on encrypted data is still rather limited. SMPC is a cryptographic protocol that allows multiple parties to contribute to a common computation without the need to show their data to the other parties. However, MPC is very resource intensive, and therefore do not scale well in in larger and more complex applications. The latter, in combination with the required domain knowledge to build such systems makes MPC unfeasible in our industry cases.

VI. Conclusions

The research which we presented in this paper is motivated by two use cases from industry partners, respectively in the context of B2B and G2B data sharing for ML purposes. The problem which we identified is that there are two major hindrances towards data-driven intelligence gathering, namely a lack of in-house ML knowledge, and insufficient data to enrich existing data sets and enabling the extraction meaningful insights. As a solution, a third-party ML expert with the necessary expertise is often brought in, which can gather additional data from public or private sources when required. However, the involvement of a third party ML party introduces its own challenges, namely that business or governmental entities now must trust these external parties with their data.

In this paper, we provide technological solutions in the form of privacy-preserving data sharing architectures to alleviate or reduce the stringent requirement of trust in a third party. We outline two architectures, depending on the degree of willingness by the data provider to share data, which is typically dictated by the sensitivity of the data involved. The first architecture involves readily-available privacy-enhancing techniques (e.g., generalization, de-identification) at the data provider-side before sharing datasets to the ML party. Optionally, a mediator can be involved such as a trusted third party to hide the source of the data set, as well as to further aggregate, mix, and generalize data sets when they originate from multiple data providers.

A second architecture is designed for situations where a data provider is highly reluctant to share data (e.g., in the case of highly sensitive data). In this case, an interface allows the ML party to present predictions from an established baseline ML model to the data provider. These predictions can be validated or used to provide feedback for improving the analytical model and deriving insights. It is crucial that even in such a case, the feedback presented does not leak any information on the original data set, which can be facilitated by means such as differential privacy. Similarly, a mediator can be involved to assume such responsibilities for a multitude of data providers.

We presented a trade-off analysis on both architectures in terms of their approach, the type of required data, and shared data, as well as the required complexity and implementation effort. The first architecture is highly feasible, although requires specific tailoring of the required privacy tactics on a per-use case basis as it is highly dependent on the quality and type of data provided. The second architecture is more visionary in its nature, with many future technological challenges that can enable validation of ML models, as well the ability to provide useful feedback to steer and improve an ML process without information leakage. This architecture provides a way to meet many of the legal requirements such as GDPR and other current and future responsibilities related to data ownership.

References

[1] The Brussels Times, *Flanders cracks down on social housing fraud*, https://www.brusselstimes.com/news/belgium-all-news/161102/flanders-cracks-down-on-social-housing-fraud, [Online; accessed 18-Aug-2022], 2021.

[2] Towards Data Science, *Smart Policing for Safer Cities: A Data-Driven Approach*, https://towardsdatascience.com/smart-policing-for-safer-cities-a-data-driven-approach-ed84e801526f, [Online; accessed 20-Oct-2022], 2020.

[3] Towards Data Science, *Predictive analytics in government decisions*, https://towardsdatascience.com/predictive-analytics-in-government-decisions-8128ba019a77, [Online; accessed 20-Oct-2022], 2019.

[4] EUR-Lex, *Regulation (EU) 2016/679 of the European Parliament and of the Council of 27 April 2016*, https://eur-lex.europa.eu/eli/reg/2016/679/oj, [Online; accessed 20-Oct-2022], 2016.

[5] L. Sweeney, "K-anonymity: A model for protecting privacy," *International Journal of Uncertainty, Fuzziness and Knowledge-Based Systems*, vol. 10, no. 05, pp. 557–570, 2002.

[6] A. Machanavajjhala, D. Kifer, J. Gehrke, and M. Venkitasubramaniam, "L-diversity: Privacy beyond k-anonymity," *ACM Transactions on Knowledge Discovery from Data (TKDD)*, vol. 1, no. 1, 3–es, 2007.

[7] European Commission, *Do the data protection rules apply to data about a company?* https://ec.europa.eu/info/law/law-topic/data-protection/reform/rules-business-and-organisations/application-regulation/do-data-protection-rules-apply-data-about-company_en, [Online; accessed 20-Oct-2022], 2017.

[8] H. Ko, "Pseudonymization of healthcare data in south korea," *Nature Medicine*, vol. 28, no. 1, pp. 15–16, 2022.

[9] R. Yacouby and D. Axman, "Probabilistic extension of precision, recall, and f1 score for more thorough evaluation of classification models," in *Proceedings of the first workshop on evaluation and comparison of NLP systems*, 2020, pp. 79–91.

[10] A. Kumar, J. Ligatti, and Y.-C. Tu, "Query monitoring and analysis for database privacy-a security automata model approach," in *International Conference on Web Information Systems Engineering*, Springer, 2015, pp. 458–472.

[11] C. Dwork and A. Roth, "The algorithmic foundations of differential privacy," *Foundations and Trends® in Theoretical Computer Science*, vol. 9, no. 3–4, pp. 211–407, 2014, ISSN: 1551-305X. DOI: 10.1561/0400000042.

[12] D. Desfontaines and B. Pejó, "Sok: Differential privacies," *Proceedings on privacy enhancing technologies*, vol. 2020, no. 2, pp. 288–313, 2020.

[13] P. Martins, L. Sousa, and A. Mariano, "A survey on fully homomorphic encryption: An engineering perspective," *ACM Computing Surveys (CSUR)*, vol. 50, no. 6, pp. 1–33, 2017.

[14] F. Prasser, J. Eicher, H. Spengler, R. Bild, and K. A. Kuhn, "Flexible data anonymization using arx—current status and challenges ahead," *Software: Practice and Experience*, vol. 50, no. 7, pp. 1277–1304, 2020.

[15] M. Sabt, M. Achemlal, and A. Bouabdallah, "Trusted execution environment: What it is, and what it is not," in *2015 IEEE Trustcom/BigDataSE/ISPA*, vol. 1, 2015, pp. 57–64. DOI: 10.1109/Trustcom.2015.357.

[16] V. Costan and S. Devadas, "Intel sgx explained," *Cryptology ePrint Archive*, 2016.

[17] N. Li, T. Li, and S. Venkatasubramanian, "T-closeness: Privacy beyond k-anonymity and l-diversity," in *2007 IEEE 23rd international conference on data engineering*, IEEE, 2006, pp. 106–115.

[18] A. Zigomitros, F. Casino, A. Solanas, and C. Patsakis, "A survey on privacy properties for data publishing of relational data," *IEEE Access*, vol. 8, pp. 51 071–51 099, 2020.

[19] G. D'Acquisto *et al.*, "Privacy by design in big data: An overview of privacy enhancing technologies in the era of big data analytics," 2015.

[20] M. M. Almasi, T. R. Siddiqui, N. Mohammed, and H. Hemmati, "The risk-utility tradeoff for data privacy models," in *2016 8th IFIP International Conference on New Technologies, Mobility and Security (NTMS)*, IEEE, 2016, pp. 1–5.

[21] K. De Boeck, J. Verdonck, M. Willocx, J. Lapon, and V. Naessens, "Dataset anonymization with purpose: A resource allocation use case," in *2021 International Symposium on Computer Science and Intelligent Controls (ISCSIC)*, IEEE, 2021, pp. 202–210.

[22] R. Hoogervorst, Y. Zhang, G. Tillem, Z. Erkin, and S. Verwer, "Solving bin-packing problems under privacy preservation: Possibilities and trade-offs," *Information Sciences*, vol. 500, pp. 203–216, 2019.

[23] C. E. Jakob, F. Kohlmayer, T. Meurers, J. J. Vehreschild, and F. Prasser, "Design and evaluation of a data anonymization pipeline to promote open science on covid-19," *Scientific data*, vol. 7, no. 1, pp. 1–10, 2020.

[24] D. Slijepčević *et al.*, "K-anonymity in practice: How generalisation and suppression affect machine learning classifiers," *Computers & Security*, vol. 111, p. 102 488, 2021.

[25] T. Carvalho and N. Moniz, "The compromise of data privacy in predictive performance," in *International Symposium on Intelligent Data Analysis*, Springer, 2021, pp. 426–438.

[26] B. Liu *et al.*, "When machine learning meets privacy: A survey and outlook," *ACM Computing Surveys (CSUR)*, vol. 54, no. 2, pp. 1–36, 2021.

[27] S. Yeom, I. Giacomelli, M. Fredrikson, and S. Jha, "Privacy risk in machine learning: Analyzing the connection to overfitting," in *2018 IEEE 31st computer security foundations symposium (CSF)*, IEEE, 2018, pp. 268–282.

[28] L. Song and P. Mittal, "Systematic evaluation of privacy risks of machine learning models," in *30th USENIX Security Symposium (USENIX Security 21)*, 2021, pp. 2615–2632.

[29] H. Hu *et al.*, "Membership inference attacks on machine learning: A survey," *ACM Computing Surveys (CSUR)*, vol. 54, no. 11s, pp. 1–37, 2022.

[30] B. Z. H. Zhao *et al.*, "On the (in) feasibility of attribute inference attacks on machine learning models," in *2021 IEEE European Symposium on Security and Privacy (EuroS&P)*, IEEE, 2021, pp. 232–251.

[31] J. Jia, B. Wang, L. Zhang, and N. Z. Gong, "Attriinfer: Inferring user attributes in online social networks using markov random fields," in *Proceedings of the 26th International Conference on World Wide Web*, 2017, pp. 1561–1569.

[32] J. Jia and N. Z. Gong, "{Attriguard}: A practical defense against attribute inference attacks via adversarial machine learning," in *27th USENIX Security Symposium (USENIX Security 18)*, 2018, pp. 513–529.

[33] J. Chen, W. H. Wang, and X. Shi, "Differential privacy protection against membership inference attack on machine learning for genomic data," in *BIOCOMPUTING 2021: Proceedings of the Pacific Symposium*, World Scientific, 2020, pp. 26–37.

[34] B. Wang and N. Z. Gong, "Stealing hyperparameters in machine learning," in *2018 IEEE symposium on security and privacy (SP)*, IEEE, 2018, pp. 36–52.

[35] H. Yu *et al.*, "Cloudleak: Large-scale deep learning models stealing through adversarial examples.," in *NDSS*, 2020.

[36] T. Lee, B. Edwards, I. Molloy, and D. Su, "Defending against neural network model stealing attacks using deceptive perturbations," in *2019 IEEE Security and Privacy Workshops (SPW)*, IEEE, 2019, pp. 43–49.

[37] M. Juuti, S. Szyller, S. Marchal, and N. Asokan, "Prada: Protecting against dnn model stealing attacks," in *2019 IEEE European Symposium on Security and Privacy (EuroS&P)*, IEEE, 2019, pp. 512–527.

[38] T. Li, A. K. Sahu, A. Talwalkar, and V. Smith, "Federated learning: Challenges, methods, and future directions," *IEEE Signal Processing Magazine*, vol. 37, no. 3, pp. 50–60, 2020.

[39] I. Dayan *et al.*, "Federated learning for predicting clinical outcomes in patients with covid-19," *Nature medicine*, vol. 27, no. 10, pp. 1735–1743, 2021.

[40] D. J. Beutel *et al.*, "Flower: A friendly federated learning research framework," *arXiv preprint arXiv:2007.14390*, 2020.

[41] N. Rodríguez-Barroso *et al.*, "Federated learning and differential privacy: Software tools analysis, the sherpa. ai fl framework and methodological guidelines for preserving data privacy," *Information Fusion*, vol. 64, pp. 270–292, 2020.

[42] Z. Sun, P. Kairouz, A. T. Suresh, and H. B. McMahan, "Can you really backdoor federated learning?" *arXiv preprint arXiv:1911.07963*, 2019.

[43] Y. Lindell, "Secure multiparty computation," *Communications of the ACM*, vol. 64, no. 1, pp. 86–96, 2020.

Towards Extracting Entity Relationship Diagrams from Unstructured Text using Natural Language Processing

Vaihunthan Vyramuthu
Department of Electronics and Computer Science
Aalen University
Aalen, Germany
email: vaihunthan@gmail.com

Gregor Grambow
Department of Electronics and Computer Science
Aalen University
Aalen, Germany
email: Gregor.Grambow@hs-aalen.de

***Abstract*—In computer science, the creation of applications usually involves the process of abstracting real world entities and relationships and creating models to be able to process these. One crucial part of this is data storage and management and therefore the creation of data models. As a first step, usually the Entity-Relationship (ER) model is used. However, the transformation from real world descriptions in natural language to standardized ER diagrams can be tedious and error-prone. Recently, Natural Language Processing (NLP) has gained much attention but this specific area is still mostly handled manually by humans. This paper describes a hybrid system for capturing ER model components from German texts using NLP. That way, time-consuming interpretation of textual database scenarios can be automated. We implemented and tested both rule-based and model-based approaches, whereas the main extraction is performed by the rule-based variant so that the entities, attributes, relationships and cardinalities can be strategically identified. The results of the model-based approach are used as a comparison to the rule-based results and can be applied for correctness checking and improvement of the results. Furthermore, we conducted a preliminary evaluation, which shows promising results. A hybrid approach can be better than a classical approach, as it combines the precision of the rule-based system with the flexibility of the model-based approach. This may lead to a more robust and reliable extraction, as errors in one of the approaches can be compensated by the other.**

***Index Terms*—Entity-Relationship Model, Natural Language Processing, Named Entity Recognition, POS-Tagging, SpaCy, LSTM**

I. Introduction

In today's data-driven world, large amounts of text are generated that can contain valuable information. Examples of this include police reports from online press portals, Twitter comments or Amazon reviews. Various analyses are possible using such text that can provide insights about future trends, user ratings and sentiment or other questions [1]. Extracting structured data from unstructured text, however, is a complex task that mostly requires manual processing. Automating this process can save time and resources and improve the efficiency of data analysis [2] [3].

At the same time, the use of database systems for storing and managing data is prevalent. A common approach to database modeling is the use of *Entity Relationship Models* (ERM), followed by conversion to a relational model and finally to *Structured Query Language* (SQL) statements to represent the real world data in a database [4].

Manual data extraction involves reading through the text and interpreting it correctly so that database modeling is implemented as efficiently as possible. The aim of this paper is to automate data extraction and thereby simplify the process of database modeling and integration. The results of this process are considered successful if the discrepancy between the human and software interpretation of a text with regard to correct ERM generation is as small as possible. This paper proposes an hybrid approach for the identification of entities and attributes, as well as the recording of entity relationships (including cardinalities). A hybrid approach may lead to increased error resistance. Rule-based systems alone are often prone to inaccurate results if the texts do not exactly match the expected patterns. The model-based approach compensates for this with its ability to learn from contexts and recognize variations. The analysis of semantic contexts and complex text structures that go beyond the recording of entities, attributes and relationships is not part of this work. The implementation of the approach focuses on the processing of texts in German language.

The paper is organized as follows: Section II presents related work, followed by Section III, containing details on the concept and architecture of the proposed approach. Implementations are presented in Section IV. Subsequently, we evaluate the implementation in Section V. Finally, the conclusion, including limitations, can be found in Section VI.

II. Related Work

In literature, there exist several approaches dealing with the creation of ER models from natural language. In the work of Ghosh et al. [5], a method is proposed that uses grammatical knowledge patterns and lexical and syntactic analyses of request texts to create ERMs. This system assumes that the input text consists only of simple subject-predicate-object sentences for correct information extraction. This sentence structure makes it possible to detect entities (subject), relationships (verb) and attributes or other related entities (object). Until the step before the domain-specific database is used to identify the ERM components, only NLP techniques, such as sentence segmentation, word separation (tokenization) or *POS-Tagging* are used [5].

The segmentation was carried out in this work in such a way that cases in English, such as "Mr. Mustermann", in which (.) appears after Mr, are not recognized as the end of the sentence. In *POS-Tagging*, the individual words are assigned to the corresponding word types, e.g., noun, verb, pronoun, adjective, preposition etc. In this publication, the recognition of entities, attributes and relationships is performed using a database and the *Support Vector Machine* (SVM) classifier [5].

Six tables (word & synonym for each ERM component) are implemented, which contain domain-specific words and synonyms. To capture related entities, the synonyms of a word from the table are given the same ID. This prevents redundant SQL tables from being created. Pronouns such as "he", "she" or "it" are already recognized in the *POS-Tagging* phase. In this phase, the pronoun is identified based on the closest previous entity that is present either in the same sentence or in the previous sentences. The technical term for this is *coreference resolution* [5] [6].

In the publication by Kashmira et al. [7], the ERM components are recorded using neural networks. Three main modules are presented, namely the preprocessing module, the machine learning module and the ER modeling module. In the first step, the preprocessing module is implemented using *NLTK* to preprocess the text. This includes steps, such as converting the text into lowercase letters, tokenization into sentences, etc. The machine learning module is then implemented using supervised learning. This module is trained with an English dataset where words are categorized into different categories including entity, sub-entity, attribute and irrelevant category. Four classifiers are taken into account when training the model: *Random Forest*, *Naive Bayes*, *Decision Table* and *Sequential minimal optimization* (SMO) [7].

To address the problem of attribute selection for entities in an ER diagram, the proposed model uses a combination of *ontology* and *web mining*. By using ontology, an attempt is made to filter relevant attributes from the extracted entities. In addition, web mining is used to obtain further information from the web that can be helpful in determining the attributes [7].

The publication by Habib [8] also follows similar preprocessing steps at the beginning, like Ghosh [5] and Kashmira et al [7]. After the parsing process, the grammatical sentence structure is obtained so that the components of the ERM can be determined based on rules. The words are converted into a parse tree structure to understand how the individual parts of the sentence are related to each other. Using appropriate rules for sentence structures, entities, attributes, cardinalities and relationships can be determined.

There are several other publications that go in a similar direction and examine the topic of automatic ERM generation in the context of NLP in more detail. One example by Omar et al. [9] describes heuristic-based analysis options for ER model generation. In contrast, Omar and Abdulla [10] pursue the approach of training a machine learning model that can extract the entities from the text. Depending on the complexity of the input text and the scope of training, the model achieves precision values of up to 85%. The results of Btoush and Hammad [11] can also be placed in a similar context. Here, a method is presented which, like [8], defines and applies certain rules for extracting information from texts.

It can be stated that two basic methodological approaches are used for ER model generation. The ERM components are determined either rule-based or using neural networks or artificial intelligence. Both approaches have advantages and disadvantages. The biggest advantage of rule-based extraction is the more time-efficient implementation as, unlike neural networks, no data preprocessing and training is required. Furthermore, the unambiguous definition of the rules ensures one hundred percent extraction probability. In neural networks, a residual inaccuracy always remains. In contrast, model-based extraction is not limited to a few rules, but can learn complex and nested sentence structures to determine the ERM components. These sentence structures can contain linguistic variations and ambiguities, which can be recognized more easily by neural networks than by fixed heuristics. The heuristic approach is more suitable for small application areas and cannot maintain its effectiveness in large application areas.

Purely rule-based methods are prone to lack of generalization, while purely model-based approaches often depend on large training datasets and have difficulties in capturing rare or complex linguistic structures. This hybrid approach has the potential to overcome these limitations by combining the strengths of both methods: The rule-based method enables accurate extraction, while the model-based method helps to validate and improve the results, resulting in a more robust and adaptable solution to different text scenarios.

III. Overview of the Architecture

Figure 1 shows a general overview of the individual processes of the proposed approach. The latter is divided into a rule-based and a model-based part. The model-based algorithm can be used either to train new or existing models or to extract the ER components from a text. However, these results are not used for the resulting ERM, but are only used to compare the rule-based results. This makes it possible to check whether the final result from the rule-based process may still need to be modified manually.

The input for both parts of the approach is a text that is saved in a *.txt* file. The *output.json* file contains all ER components and relationships found in a structure that can be read by an external ER modeling tool. The artifacts (in the image: ordinary rectangles or arrow labels) represent the created files or results. These files are required for the subsequent processes.

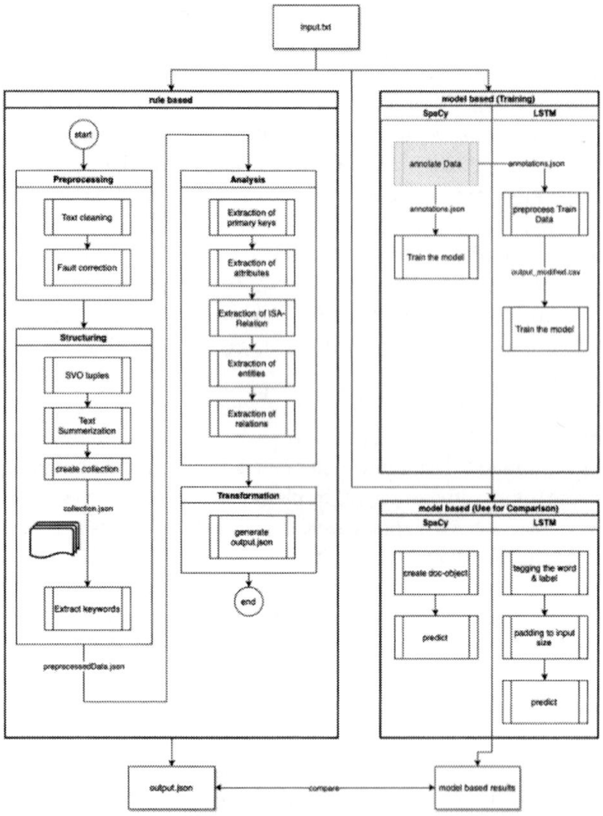

Fig. 1. Overall flow chart of the implementation.

The rule-based approach is made up of the sub-steps pre-processing, structuring, analysis and transformation. During preprocessing, the text data is cleansed and corrected for spelling errors. The structuring phase contains processes that break down the text into more meaningful parts and store them temporarily. In the main block of the analysis, the ER components are extracted one after the other.

In the model-based part, a different process path is carried out depending on the selected model type (*SpaCy*-Transformer or *LSTM*). Due to the individual input requirements of the models, the text data must be converted into the permitted form for both the training case and the use case. The process block highlighted in yellow indicates an annotation process that is carried out using an external tool. For the *LSTM* model, the output of the annotation tool must be pre-processed again into a *.csv* file so that it can be used for training. After the training processes, the model can be applied to new text data. For the *SpaCy* model, it is sufficient to convert the text into a doc object so that the analysis can be started. The text for the *LSTM* model, on the other hand, requires additional processing (analogous to the training process), which decodes the words and labels into numbers and scales the input word vector to a specified size (so-called *padding*). This hybrid combination provides an important advantage for continuous optimization. The results of the model-based approach can be used to dynamically adapt and further develop the rules of the rule-based system. This means that the hybrid system can become increasingly precise and effective over time through feedback and new data sets.

The selection of *SpaCy* as a model and NLP-Tool is based on its robust capabilities and user-friendly implementation. SpaCy is recognized for its comprehensive documentation and strong support from an active developer community. *LSTM* were chosen for their effectiveness in handling sequential data and their ability to capture contextual dependencies over extended text passages. This makes them particularly suitable for tasks that require understanding the relationships within long text sequences of information.

IV. Implementation

Starting with the rule-based part, it can be noted that it is important not to define too many special rules. Otherwise, the implementation will be too application-specific and errors for other text styles will sometimes occur. The first process in data preprocessing is text cleansing. This process is divided into three steps. Each result of the individual preprocessing and structuring steps is saved in the *preprocessedData.json*. At the beginning, the unstructured text is filtered from the *.txt* file.

In this first step, existing white-spaces or empty lines are also taken into account. In the second step of text cleansing, the individual sentences from the text are found and saved using the *SpaCy* model (*de_core_news_sm*). In the last step of the text cleansing process, unnecessary sentences that do not provide the necessary information for the ER diagram design are removed. In this process, sentences are removed using *RegEx* matches of certain words, such as "database" or "modeling", are sorted out.

The next process in preprocessing is the error correction of words. The *Levenshtein-similarity* is used here. The Python library *pyspellchecker* checks whether there is an error for each word in a sentence. If this is the case, the word is replaced with the closest one.

In the structuring task block, the main focus is on capturing the subject-verb-object (SVO) sentence structure and some important term frequency analyses, with the help of which the text can be broken down into smaller information-rich parts. While only the *normalized word frequency* (TF) is used for the text summary, the *inverted document frequency* (TF-IDF) helps to capture the most significant keywords from the text. In contrast to SVO generation, these two structuring steps are only used optionally. These results are not actively used in the NLP workflow, because it is possible that relevant information may be lost. Another use case for these results is looking for the most essential keywords in the text in order to compare them with the entities and attributes found. It should also be noted that to calculate the TF-IDF for keyword extraction, a document corpus (*collection.json*) must be created in which all existing ER diagram sample texts are stored. The *TfidfVectorizer()* function provided by the *Scikit-*

learn library calculates the TF-IDF. No numbers are included in the keywords.

The text summary is implemented chronologically according to the following key points:

- count the number per word in the text (stop words excluded)
- calculate the normalized weight per word used by dividing the respective word count by the maximum word frequency occurring.
- calculate the sentence weight by adding the weight per word.
- search for sentences with the highest weighting.

The individual sentences are scored by adding the normalized TF for each word in the sentence. Depending on the original length of the text, a certain number of sentences is selected in descending order of the evaluation number. For this purpose, a corresponding factor is determined at the beginning, with which the number of sentences is calculated. If there are fewer than three sentences in the text, the text is not summarized.

To generate SVO tuples, the sentence must be analyzed using *dependency parsing* and *POS-Tagging* provided in *SpaCy*. Certain commands can be used to analyse the grammatical structure of sentences so that the visualization shown in Figure 2 is displayed. Similar to the tree structure, the successors or predecessors of a word are addressed with "children" or "parent".

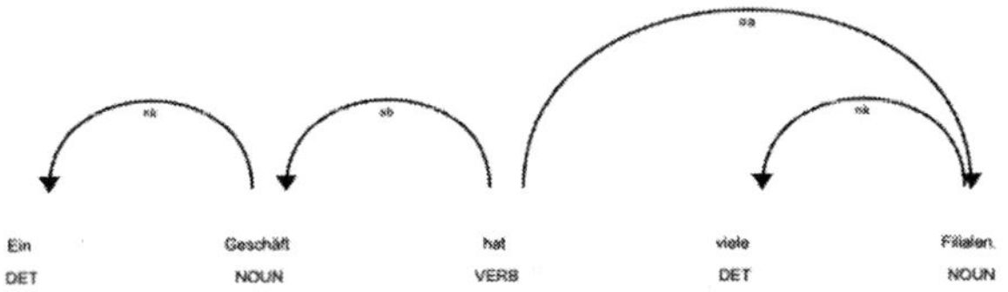

Fig. 2. Visualization of the grammatical sentence structure.

On the basis of this structure, a generally valid logic can be developed for the extraction of subject, verb and object, which is shown in the structure diagram in Figure 3. The sentences from the text are entered individually into the function. There may be several verbs in each sentence, but each verb must belong to exactly one subject and object. *Lemmatization* can be used for the output of recorded relationships in texts. This involves changing the verb from its inflected form back to its basic form. For example, the German word "angeboten" is "anbieten" after lemmatization.

A. Primary Key

The primary keys can be found using a *RegEx* comparison. The words "-id" or "-number" indicate a key candidate.

The words are first converted to lower case so that there is a certain amount of leeway in the comparison. However, the limitation is the hyphen, which must be contained in a primary keyword. The following rules are implemented as Python code:

- Determine all nouns and filter primary key via *RegEx*.

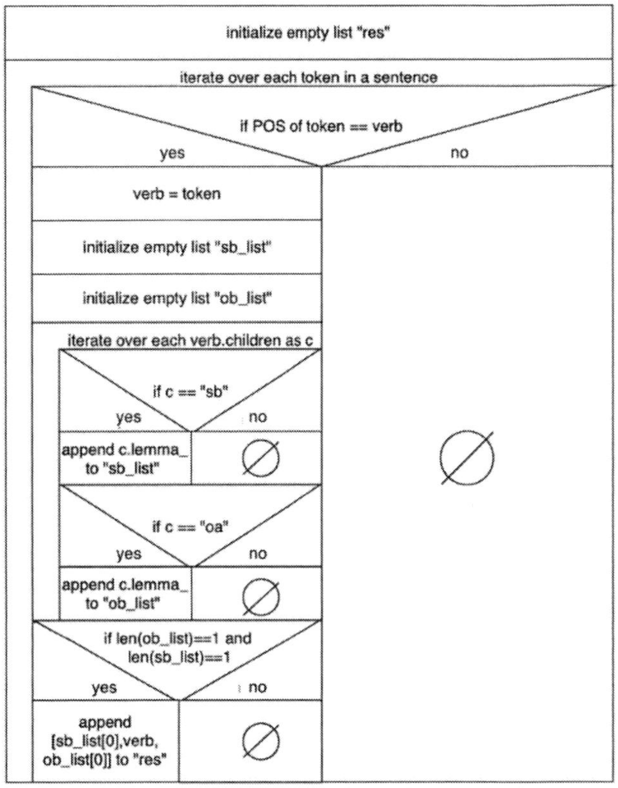

Fig. 3. Structogram for SVO extraction from a record.

- If the record only contains the word "unique", then the closest noun should be the primary key with "noun ID".
- If only the words "ID" or "id" appear, then the nearest noun should be the primary key with "noun ID".

B. Attribute

Attributes are identified as soon as a sentence contains a list of more than two nouns. The first noun that occurs in the sentence is the entity to which the remaining nouns or attributes belong. This simple rule serves as a first step in the identification process, but it can be refined and enhanced through model-based results in the future.

C. ISA-Inheritance

To detect the specialization or generalization of entities, the following rules are followed:

- If a sentence contains the following verbs: ["include", "consist", "comprise", "share", "include"], then the sentence describes a generalization.
- If the sentence contains "type" as a word (noun), the first noun or entity is the generalization of the following nouns (entities).

It should be noted that the recognition of ISA relationships using rule-based approaches is limited. For example, the order of membership may be different for the entities or the ISA description may extend over several sentences.

D. Entity

In the *preprocessed.json* under the item in which the SVO tuples are stored, the subject and object in each record can be either an entity or an attribute. These tuple words are therefore compared again with the attributes and primary keys found so far. If the same tuple word is also contained in the list for attributes or primary keys, it is discarded. The final result is a list that only contains the final entities.

E. Relationship

The SVO tuples can again be used for the relationships. If both the subject and the object are contained in the list of final entities, the verb is a relationship between two entities. There are sentences that are not formulated in an ordinary SVO style, but which define further relations between entities. In order to take such sentences into account as well, a check is made to see whether two nouns occur in a sentence in addition to a verb, which are not contained in the attribute, ISA-Inheritance and primary key list. If this is the case, a relationship tuple can be extracted from this sentence again if it has not already been extracted from the SVO tuple.

F. Cardinality

Two min/max cardinalities must be determined for each relationship between two entities. The sentence in Figure 2 shows that the cardinalities can be taken from the determiners of the nouns. Once the SVO tuples have been reassigned to the complete sentences with the help of indexing, the corresponding determiners of each entity can be determined. These are then translated into the corresponding min/max value using a comparison. For the correct min/max notation, the cardinalities of the entities in an SVO tuple must be swapped.

When interpreting the adverbs “at most” and “at least”, the following word must also be taken into account, as this defines either the upper (max) or lower limit (min).

Due to the unlimited possibilities for translating this adverb, no rule-based application is suitable for this. This makes the rules too specialized for one use case.

The model-based approach primarily serves as a comparison tool for the ER components found in the rule-based algorithm. Therefore, the hybrid approach is also advantageous for performance reasons. The rule-based methods often deliver faster results as they do not rely on extensive calculations, while the model-based part intervenes where more in-depth analyses are required. This efficient applicability ensures that the goal is achieved faster without losing accuracy.

V. Limitations

The following aspects were not taken into account in the rule-based algorithm:

- Attributes for relationships.
- special cardinalities (“two”, “three”, etc.).
- weak entities, relationships, attributes, etc.
- described ISA-Inheritance across several sentences.
- multi-valued or complex attributes.

In the future, results from the model based approach could be used to include more ER components in the results. An extension to the rule-based approach for this would be rather difficult, as a generally valid formulation of the rules is difficult and often tied to a specific use case.

VI. Evaluation

The evaluation is based on several German texts that describe a specific DB scenario. This means that they contain specific formulations that describe the ER components. The rule-based algorithm only extracts ER components that correspond to the defined grammatical regularities. In contrast to the model-based approach, emphasis is placed on a qualitatively correct ER extraction instead of a quantitative result set.

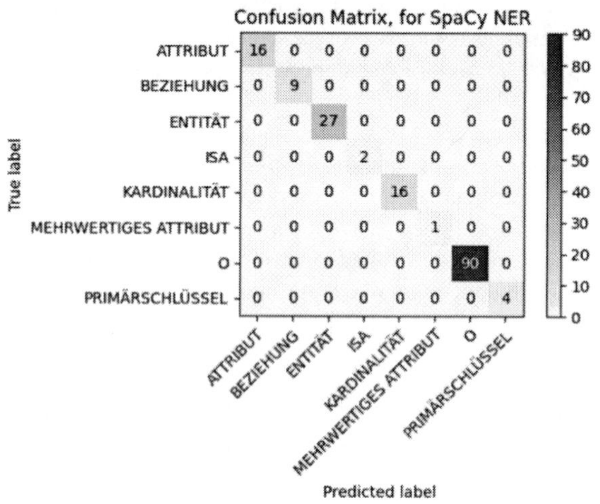

Fig. 4. Ideal reclassification result in the case of training and testing with the same data.

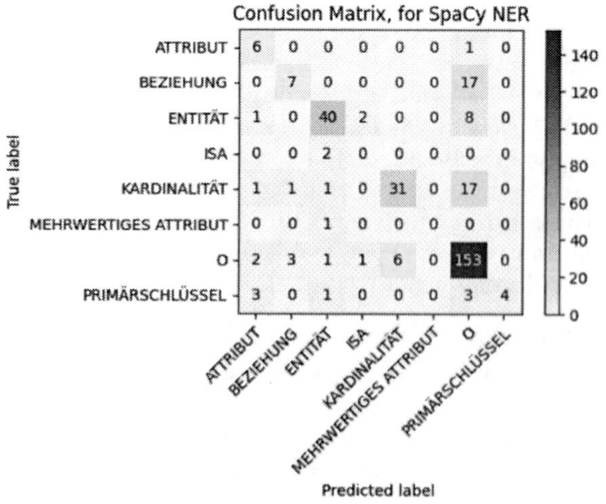

Fig. 5. Results of Crossvalidation with larger and different training and test data sets.

The Confusion Matrix in Figure 4 shows the reclassification case in which all previously labeled ER components are correctly classified during testing. However, the Confusion Matrix in Figure 5 provides more information about the generalization capability of the *SpaCy* model, because the cross-validation also examines text data that the model does not yet know and was therefore not part of the training process.

Analogous to the *SpaCy* model, the *LSTM* model is also evaluated using the learning curves shown in Figure 6. After 10 training epochs, a solid validation accuracy of 92.1% and a validation error of 0.53 are achieved.

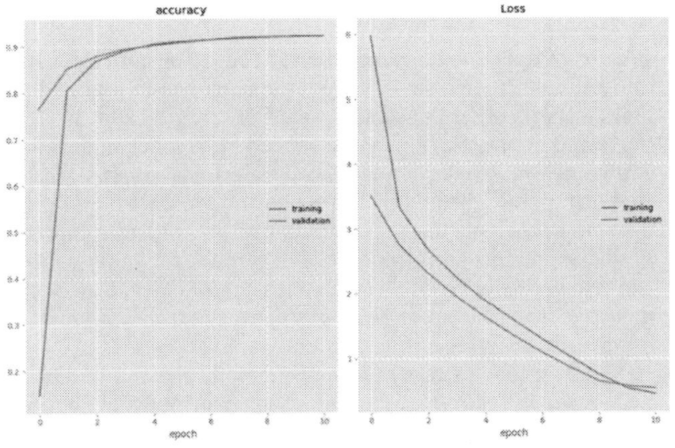

Fig. 6. Learning curves from the *LSTM* model training.

This can only serve as a first step for an evaluation. As part of our future work, we plan to compare our results to the results created by domain experts for a set of different models.

The results of the two approaches, both rule-based and model-based, complement each other very well to form a complete tool due to their complementarity. It can be stated that the challenge of the rule-based method lies in the assignment of the nouns found to the individual components. Furthermore, the connection of attributes and relationships to certain entities was not entirely trivial. On the other hand, the model-based comparison offers possibilities to correctly capture different cardinalities or ISA inheritances due to the increased flexibility in sentence structure and grammar structure.

VII. Conclusion

Extracting information from unstructured text is a complex task. Manual processing of large amounts of text is time-consuming, error-prone and not scalable. Recently, numerous approaches for automating this task have been proposed. However, there exist many cases where very specific information has to be extracted from unstructured text. These imply challenges based on the specifics and rules applicable for the desired result. One of these cases is the extraction of ER models. There exist several approaches, but they still lack different features for a complete and usable automation.

We proposed a hybrid approach to extract meaningful ER data from unstructured text. Two subsystems were developed, both of which can extract ER components from unstructured texts. Special NLP methods were used for the rule-based extraction of entities, attributes and relationships. Due to the higher reliability of the rule-based results, these were used for the final ER model. *SpaCy* and *LSTM* models can be used to validate the rule-based results. In the future, the results of the rule-based approach could be supplemented by an automated comparison of the results from the model-based approach.

In literature, there are still research gaps in the area of automatic extraction of ER models from texts using NLP techniques. This includes the integration of contextual information and the consideration of ambiguity. Another fundamental improvement is the training of the models with even more text data, so that even rarely occurring ER components, such as multi-value attributes, can be better learned. The pretrained models can be trained for other domains according to the principle of transfer learning so that the models can be used for other purposes, e.g., to create knowledge graphs.

Our future work will include the mentioned gaps: We will expand our approach to include more specific ER concepts. Further, we will concentrate on enabling a broad applicability for different domains. We will also investigate options for automated integration of the results from the two approaches. As a first step, we will extend our evaluation including use cases with real world texts and compare the results of our approach with ER models created by human experts.

References

[1] M. Kanakaraj and R. M. R. Guddeti, "Performance analysis of ensemble methods on twitter sentiment analysis using nlp techniques," in *Proceedings of the 2015 IEEE 9th international conference on semantic computing (IEEE ICSC 2015)*. IEEE, 2015, pp. 169–170.

[2] A. Rajput, "Natural language processing, sentiment analysis, and clinical analytics," in *Innov in health informatics*. Elsevier, 2020, pp. 79–97.

[3] R. Suganya, R. Krupasree, S. Gokulraj, and B. Abinesh, "Product review analysis by web scraping using nlp," in *Smart Data Intelligence: Proceedings of ICSMDI 2022*. Springer, 2022, pp. 427–436.

[4] V. T. N. Chau and S. Chittayasothorn, "A bitemporal sql database design method from the enhanced entity-relationship model," in *2021 7th International Conference on Engineering, Applied Sciences and Technology (ICEAST)*. IEEE, 2021, pp. 85–90.

[5] S. Ghosh, P. Mukherjee, B. Chakraborty, and R. Bashar, "Automated generation of er diagram from a given text in natural language," in *2018 Int'l Conf on ML and Data Eng (iCMLDE)*. IEEE, 2018, pp. 91–96.

[6] V. Bryl, C. Giuliano, L. Serafini, and K. Tymoshenko, "Supporting natural language processing with background knowledge: Coreference resolution case," in *The Semantic Web–ISWC 2010: 9th International Semantic Web Conference, ISWC 2010, Shanghai, China, November 7-11, 2010, Revised Selected Papers, Part I 9*. Springer, 2010, pp. 80–95.

[7] P. Kashmira and S. Sumathipala, "Generating entity relationship diagram from requirement specification based on nlp," in *2018 3rd Int'l Conf on Information Technology Research (ICITR)*. IEEE, 2018, pp. 1–4.

[8] M. K. Habib, "On the automated entity-relationship and schema design by natural language processing," *Int'l J Eng Sci*, vol. 8, no. 11, pp. 42–48, 2019.

[9] N. Omar, J. Hanna, and P. McKevitt, "Heuristic-based entity-relationship modelling through natural language processing," in *Proc. of the 15th Artificial Intelligence and Cognitive Science Conference (AICS-04)*. Artificial Intelligence Association of Ireland, 2004, pp. 302–313.

[10] M. Omar and A. Abdulla, "The entities extraction for entity relationship models from natural language text via machine learning algorithms," in *Proc 4th Int'l Conf of Basic Sci and Their Appl, Elbeida, Libya*, 2020.

[11] E. S. Btoush and M. M. Hammad, "Generating er diagrams from requirement specifications based on natural language processing," *Int'l J of Database Theory and Application*, vol. 8, no. 2, pp. 61–70, 2015.

Evolving the Automated Search for Clusters of Similar Trajectory Groups

Friedemann Schwenkreis
Business Information Systems
Baden-Wuerttemberg Cooperative State University
Stuttgart, Germany
email: friedemann.schwenkreis@dhbw-stuttgart.de
ORCID: 0000-0003-4072-0582

Abstract— **The work presented in this paper builds upon a previous approach to automatically detect tactics based on spatiotemporal data in the context of team handball. It will be shown how the availability of additional data allows us to verify the principal approach. However, it will also be shown that the previous approach for choosing parameters of the applied methods was suboptimal, and an application-oriented approach based on heuristics helps to improve the results significantly. Basically, the combination of Shared Nearest Neighbor Clustering and the search for frequent itemsets is used to find clusters of trajectory groups. These basic methods are enhanced by special notions of distance and cluster quality indexes which allows to find optimal parameter settings for the specific application scenario. Furthermore, an approach is presented to use the existing "composite model" to determine the cluster to which a group of trajectories belongs to (application of the composite model).**

Keywords- trajectory sets; SNN clustering; frequent itemsets; tactics recognition.

I. Introduction

Previous work proposed using data from position tracking systems, such as Vector from Catapult or from Perform LPS by Kinexon to automatically process the position data of players of team ball games [1][2]. Schwenkreis proposed a deep learning-based classification approach to automatically recognize team tactics based on the abovementioned spatiotemporal sensor data [3]. The approach was subsequently modified to avoid the large amount of necessary training data and thus the need for labeling data [4]. The proposed solution was to use clustering based on the Fréchet distance [5] of trajectories combined with a silhouette coefficient-based [6] quality criterion to cope with noise. A subsequent paper enhanced the approach by avoiding the shortcomings of the Fréchet distance and eliminating the need for generating ordered sets [7]. Furthermore, the enhancement avoids the need for a distance criterion of sets of trajectories by introducing a combination of clustering and the search for frequent itemsets.

In his latest paper, Schwenkreis explicitly identified the need to collect additional trajectory data to extract a stable set of groups of trajectory sets. However, there was no discussion regarding how to incrementally improve the model or how to determine a stable state. Furthermore, the actual objective of identifying frequent sets of trajectories is to identify tactical patterns that can later be used to automatically detect them in streams of trajectories. Hence, there needs to be a mechanism to decide whether a set of trajectories belongs to one of the previously identified clusters.

Based on the previously introduced basic mechanisms for extracting a cluster model of trajectory groups from spatiotemporal data, this paper will present the latest developments of the extraction of a cluster model. Furthermore, this paper presents an approach for applying the extracted cluster model to determine whether trajectory groups extracted from a stream of trajectory sets belong to one of the previously identified clusters.

An overview of related work is given in Section II. Then, Section III will introduce the underlying data model in Section III.A and the general clustering approach in Section III.B. An abstraction is introduced that allows the application of the approach in arbitrary situations in which similarity clusters are detected in sets of trajectory groups. In Section IV, the necessary distance functions, the quality criteria for clusters, and the assignment to clusters are discussed. Section V presents the results of an evaluation based on a real-world application of the approach. The paper is concluded with a summary and an outlook on the future application of the approach in Section VI.

II. Related Work

A. Pattern Recognition

Pattern recognition in the context of spatiotemporal data has a long history [8], and a significant number of related studies have been published in the area of trajectory clustering. Trajectory clustering provides the foundation for recognizing patterns in sets of team moves. However, since team moves consist of groups of trajectories, the targeted problem of tactic recognition is not identical to the family of problems that is addressed by classical trajectory clustering. Nevertheless, recent trajectory clustering approaches, such as those described in [9], have made significant progress in the area of trajectory clustering, and the work presented in this paper has been influenced by these approaches.

In particular, the flock pattern and later generalizations to the convoy pattern seem to overlap with the problem addressed by this work [10]. Unfortunately, there are significant differences that prevent the direct application of these approaches:

- Current convoy mining approaches assume that clusters are searched in sets of points that have been collected at the same point in time. This is not the case for the class of problems discussed in this paper. In contrast, the task is to search for clusters of trajectories that have not(!) "happened" at the same point in time. Furthermore, there is no fixed mapping on a common logical clock that would allow us to treat the coordinates of trajectories as if they were collected at the same time.

- The usual approaches to trajectory clustering are based on the notion of density and use a density threshold to determine the clusters. The basic assumption of these approaches is that a single density threshold can be found, which is not possible in the given case. By inspecting the application area, it is known that the density of the trajectories significantly differs across different trajectory clusters (see also Section IV.B).
- Convoy mining approaches assume that points of trajectories belonging to the same cluster also belong to a single cluster. In the given application scenario, this does not need to be the case. In terms of convoy mining approaches, points of the same trajectory may join other convoys and return to the original convoy because trajectories belonging to different clusters may have non-empty intersections from a geometrical point of view because they have identical points.

The field of detecting optimal care pathways in health care [11] has some similarities to detecting team tactics based on an abstract notion of trajectories. Clustering approaches in this area are based on a completely different notion of distance [12]. Furthermore, the requirements regarding the temporal distance of "locations" differ significantly, which leads to methods that are based on the sequence only rather than considering the real distances in time. Hence, the work in that area cannot be applied in the given context.

B. Sports Analytics

Recently, the analysis of spatiotemporal data in the context of sports has attracted increasing attention. There are several attempts in the area of team sports to exploit the data that are produced by the position sensors carried by players [13]- [15]. Several activities focus particularly on soccer (or football) to extract models that help to explain the mechanics of the game [16]. This is because professional soccer teams can fund analysis projects, and there is still no accurate model for computing appropriate predictions. Some work can be found in the literature that derives patterns from spatiotemporal data, but these approaches use classification to predict, for example, ball losses or scoring probabilities because in these cases, the target value is available in the automatically collected data. This is not the case for tactical labels but essential for the case presented in this paper.

Unfortunately, the mentioned work does not focus on detecting patterns of moves of groups of players. The reason is that soccer and other team sports significantly differ from team handball in terms of the speed of attacks. In the case of team handball, it is crucial that the individual moves (and resulting positions) of teammates are known upfront by the other players because, in most cases, the determination of the location of other team members by an explicit visual observation is too slow. Thus, the coordination of the players' moves is trained based on explicitly communicated movement patterns (called tactics). Ice hockey has some similarities to team handball because the players' speed is even greater than that of team handball players. However, tactics are focused on the movement patterns of individual players rather than on the coordinated patterns of a whole team, which has also been reflected in recent work that targets the analysis of spatiotemporal data in the context of ice hockey [17].

III. BASICS

A. Underlying Data Model

1) Individual trajectories

As proposed in the aforementioned previous work, the 2D coordinates delivered by tracking sensors are "normalized" to avoid differences due to changing directions of play [4]. Given an observation interval of t_2-t_1 seconds and a position sampling rate f, the individual trajectory $T_s(t_1,t_2)$ of a sensor is defined as the timely ordered set of $r=(t_2-t_1)*f$ coordinate pairs p: $T_s(t_1,t_2)=(p_1, p_2,\ldots ,p_r)$. It is assumed that all the observed sensors generate samples at the same rate in the given application context. Furthermore, there is a mapping for individual trajectories T_s that assigns a team O to the individual trajectory: $team(T_s)$: $T_s \rightarrow O$.

2) Team Moves: Sets of trajectory class identifiers

The notion of a team position has been defined by Schwenkreis as a vector of positions of the contained sensors (team members), which is similar to the coordinate of a team in a $2n$-dimensional space, where n denotes the number of sensors [3]. The challenge of this approach is to associate a specific sensor with a well-defined position in the vector. This approach is challenging because the collection of sensors that comprises a team is not constant and might change due to the substitutions of team members. Originally, all possible permutations of a vector were generated when training a model (which is rather costly) [3]. Generating permutations is avoided in subsequent work by introducing a so-called canonical sort order of sensor positions contained in a team position [4]. The sorting order is based on additional information regarding the individuals who are carrying a sensor. As a result, there is a unique mapping of sensors to positions in a team vector.

Alternatively, to the approach above, a core assumption regarding individual trajectories can be exploited. Individual trajectories are not randomly distributed across the feature space. There are rather clusters of similar trajectories that are intentionally followed. Hence, there exist only a limited number of trajectory classes, each representing an *intended trajectory* given a certain context. In application terms, this can be called the *intended individual contribution* given an intended team tactical move. Based on this assumption, individual trajectories $T_s(t_1,t_2)$ can be mapped onto identifiers of intended trajectory classes: $T \rightarrow c$, $c \in \mathbb{N}$. There might be cases where individual trajectories do not match with any intended trajectory class. In these cases, the individual trajectory is mapped onto the "noise" class identifier represented by a value of *-1*. As a result, a team tactical move M of a time interval can be defined as a tuple of trajectory class identifiers of the time interval $M(t_1,t_2)=(c_1, c_2, ..., c_n)$ with n in the range of one to the number of sensors belonging to the observed collection, also called the team or group size. The sorting order of the class identifiers contained in M is irrelevant. Thus, we simply assume that the class identifiers are given in descending order: $\forall c_i, c_k \in M : i<k \rightarrow c_i >= c_k$.

B. Clustering Aspects

1) Two-Step Approach

The automated detection of tactics based on clustering was proposed in [4] to avoid the need for labeling data. The approach did not explicitly select a clustering technique but introduced a quality criterion to compare different techniques. The mentioned approach uses spatiotemporal data that comprise groups of trajectories of team members to search for similar team moves by clustering. In a given example application scenario (team handball), this results in records of 1.760 attributes (880 pairs of 2-d coordinates) per team move [4].

This number of attributes is rather high, and a number of clustering approaches have been described in the literature to reduce the dimensionality of the data, particularly in the context of time series data (such as trajectories) and [18]. A special group of these approaches is the set of multilevel or multistep clustering methods, which (from a high-level perspective) follow a stepwise approach to reduce the dimensionality of the data by clustering a "sub-aspect" first. The sub-clusters are then clustered again on the next level. For example, Aghabozorgi et al. introduced a two-step clustering approach for time series data by starting with a fine-granularity cluster search, which is followed by a subsequent clustering step to merge similar clusters using a different criterion that is specific for the next level [19].

The basic idea of two-step clustering is adopted for the case of this paper to address the dimensionality challenge. Rather than directly trying to find clusters in a set of trajectory groups, it is proposed to search for clusters at the trajectory level first, given a trajectory-specific distance criterion. Subsequently, a search for similarity clusters of trajectory group clusters (denoted as team tactics) is performed. Thus, rather than having to find clusters based on $2nk$ attributes (when n is the number of positions of the trajectories and k is the number of trajectories per group), the clustering of the first step has to find clusters in records with only $2n$ attributes. The subsequent step has to handle records consisting of only k attributes. Projected on the application case of [4], there are only 220 rather than 1.540 attributes on level one and 7 attributes on level two (each representing a the individual move of player of a team).

The two-step approach is particularly promising for trajectory groups because it usually provides a meaningful explanation on the application level. The first clustering searches for patterns of individual contributions, while the second search focuses on patterns of combinations of individual efforts. In application terms from team ball games: what are the intended moves of players (or player types), and how are team tactics composed of these individual moves?

2) Clustering of trajectories

There is always "noise" in the trajectory data in the given context because there will always be player moves that are not intended moves in the sense of a contribution to some tactics. Thus, only clustering techniques can be used that can cope explicitly with noise, which excludes, for instance, basic spectral clustering [20]. Even later enhancements of spectral clustering called robust spectral clustering can only handle a low number of noise points compared to the number of non-noise points [21]. Furthermore, no assumption regarding the cluster shape can be made. Trajectory clusters might have concave boundaries, which excludes clustering techniques, such as k-means clustering. Based on this, only two clustering concepts have been further investigated: agglomerative hierarchical clustering [22] and density-based spatial clustering of applications with noise (a.k.a. DBSCAN) [23]. However, agglomerative hierarchical clustering can be simulated using particular parameter settings of DBSCAN. Thus, this paper focuses on DBSCAN only.

3) Finding similarity groups of team moves

As described in Section III.A, team moves are represented by ordered *k-tuples* of trajectory cluster identifiers. To find groups of similar team moves, another clustering step can be used, but we lack a meaningful notion of distance for team moves. A straightforward approach would be to use the Hamming distance (based on [24]), as in the case of distances of words, which has no meaningful interpretation in the context of team moves.

Alternatively, the search for similarity groups can be performed based on the method of searching for *frequent itemsets,* as is done in the case of association rule mining [25]. With this approach, all *frequently occurring* combinations of previously extracted cluster identifier combinations will be found without the need for a distance or similarity criterion. However, not every previously identified trajectory cluster is relevant when identifying team tactics. For instance, there are clusters that represent player trajectories in which the players (almost) do not move at all. These trajectory clusters can be seen as *passive* contributions to a team tactic rather than *active* contributions. Consequently, the trajectory clusters must be *weighted* based on the distance covered by the contained trajectories to reflect the contribution to a team tactic.

When weighting trajectory clusters, the search for frequently occurring clusters needs to take weights into account, which is comparable to the process of searching weighted itemsets. In previous work in the area of weighted itemsets, the weights became somewhat part of the notion of frequency [26]. That is, the low weight of an itemset can be "compensated" by high support to still have a frequent itemset and vice versa. In the given application scenario, this is not the case. A trajectory cluster with a low weight is considered to be of low relevance regardless from its frequency. Even a high support of the cluster will not "make it more relevant". In application terms, if a player does not move, there is no relevance of the trajectory with respect to a team tactic, no matter how often this occurs.

The weight of a trajectory cluster is defined as the length of the trajectory (sum of the Euclidean distances of the contained points) representing the containing cluster t_r^c. The representative trajectory of a cluster is defined as the trajectory with the minimal distance to all other trajectories of the same cluster: $t_r^c = t_i | \forall\, t_i, t_k \in c$: $D_i^c \leq D_k^c$ and D_i is the sum of all distances of a trajectory of a cluster to any other trajectory of the same cluster $D_i^c = \sum dist(t_i, t_k) \mid t_i, t_k \in c$.

The *relevance coefficient* r_i is assigned to the tuples t_i representing team moves: $t_i \rightarrow r_i$, $r_i \in \mathbb{N}_0$. The relevance coefficient represents the number of contained trajectory cluster identifiers that identify a cluster whose representative trajectory t_r^c has a length greater than a specified threshold. The search for *relevant frequent itemsets* identifies the sets of trajectory cluster identifiers that have a support s_i greater than a given minimum support and a relevance greater than a given threshold: $\{c_i\} \mid s_i > s_{\min} \wedge r_i > r_{min}$.

The Apriori approach to finding frequent itemsets [25] can be easily extended to cover cases with a relevance coefficient. The basic idea of Apriori is that the support of an itemset containing a certain number of items cannot be greater than the support of any subset containing fewer items. The relevance coefficients of team moves do not have this property in general because the relevance coefficient of an itemset cannot exceed the number of contained items. Hence, the straightforward approach is to use regular Apriori to generate the frequent itemsets, which are subsequently checked for their relevance based on the assigned relevance coefficient of the contained trajectory clusters and the specified minimum relevance. After the identification of the relevant frequent itemsets, itemsets containing non-relevant items can be eliminated to focus on team moves with relevant trajectories only.

The straight-forward approach can be improved by using the relevance as a sort criterion of the items contained in an itemset (the itemset becomes a tuple). As introduced in [26], itemsets can be treated as sorted sets (tuples) based on the decreasing relevance of the contained items. The candidate generation then combines only trajectory cluster identifiers that represent a cluster whose representing trajectory has a length greater than the specified threshold (which means a contribution to the relevance coefficient greater than zero), and candidate itemsets with nonrelevant items are *pruned*. Finally, the resulting itemsets need to be checked for the minimum relevance limit and support.

4) *Assigning team moves to itemsets*

To group the team moves, a mapping of each team move t_i onto one identified relevant frequent itemset f_k is needed: $t_i \rightarrow f_k$. A naïve approach would be to directly assign an itemset to any team move that supports the itemset. Unfortunately, this simple association is ambiguous because itemsets can have a subset relationship, and a single team move might even support multiple itemsets not having a subset/superset relationship. The latter case is an indication of not having enough data to be able to identify the "missing" superset of the union of the supported itemsets as relevant and frequent. The association of a team move to any of the itemsets can be chosen arbitrarily in this case. The case of nested itemsets is rather simple. A team move should be associated with the relevant frequent itemset that consists of the maximum number of items that is supported by the team move. It represents the specialization of another tactical move—the subset.

IV. Distances, Similarity, and Quality Indicators

A. *Trajectory Distance*

A distance or similarity function for individual trajectories is needed to be able to find clusters of similar trajectories in the absence of a labeling attribute (the ground truth) in the data. In previous work, the discrete Fréchet distance [5] was used as the distance between two trajectories without an in-depth discussion of alternatives. In a more recently published comparison of trajectory distances, it was shown that the discrete Fréchet distance is sensitive to outliers and to timely shifts in trajectories [27]. It is also shown that dynamic time warping [28] outperforms the Fréchet distance in scenarios that are similar to the scenario addressed by this paper. However, dynamic time warping is not a metric (missing the triangle inequality property), which limits its applicability. Fortunately, the used approaches do not rely on the triangle inequality property because of their independence from path-length based criteria.

Continuous dynamic time warping was not covered by the comparison but was identified in later work as the most flexible distance criterion for trajectory distances in [29]. Continuous dynamic time warping was derived from dynamic time warping to cover cases in which the discretizations of trajectories are not uniform. However, in the cases addressed in this paper, uniform discretization can be guaranteed because the individual trajectories consist of the same number of coordinates distributed evenly over time. Thus, continuous dynamic time warping has no advantage compared to the original dynamic time warping approach in the given scenario. Paparrizos et al. argue against the use of dynamic time warping when clustering time series data [30]. However, this process is performed based on a generalized case without considering specific cases, such as a time series of two-dimensional position data. The concerns raised in the paper do not hold in this specific case.

The advantage of dynamic time warping compared to the discrete Fréchet distance is the concept of "warping", which tolerates time shifts between the coordinates of two trajectories. Furthermore, the discrete Fréchet distance focuses on the maximum distance between pairs of coordinates, while the dynamic time warping distance is the sum of all distances between matching pairs, thus smoothing the effect of outliers. In conclusion, dynamic time warping is the optimal method for determining trajectory distances in the context of this work. However, the dynamic time warping distance has been slightly adapted to avoid a dependency on the number of points a trajectory consists of. Rather than the sum of all distances, the average distance is used.

The optimal warping window size is highly application dependent. It depends on the accuracy of the frequency of the position detection technology as well as the absolute speed of the sensors. Furthermore, the notion of similarity of a given context limits the time gap that is tolerated when two trajectories are compared. In case of team handball moves, the time gap must be in the sub-second range. Non-comprehensive experiments with a tolerable time gap of up to half a second have shown acceptable results.

B. Shared Nearest Neighbor Trajectory Similarity

Distance-based clustering algorithms, such as DBSCAN, look for dense areas based on a single distance-based threshold. Consequently, algorithms cannot cope well with areas with varying densities. In the case of varying densities, clusters with lower density are not found if the distance threshold is set too low. On the other hand, if the distance threshold is too high, then multiple clusters with high density might be merged, and additional details may be lost.

Unfortunately, the application scenario of this work must explicitly handle varying densities because the running distances of different player roles (positions in a team) and thus the DTW distances differ significantly. The so-called *shared nearest neighbor similarity* is an approach for handling varying densities while still using the original notion of distance as the underlying criterion [31]. Shared nearest neighbor similarity uses a notion of similarity that depends only indirectly on distance. Conceptually, the approach computes a list of nearest neighbors for each record based on the chosen notion of distance (the dynamic time warping distance in the case of this paper). When the similarity of two records is computed, prefixes of length l (a user-specified limit) of the records' lists of nearest neighbors are compared. The number of nearest neighbors contained in both lists is the value for the similarity of the two records. Then, a DBSCAN-like clustering approach searches for clusters based on similarity values.

The notion of similarity has a significant limitation: the similarity value depends on the number of points that are compared. Thus, the notion of similarity has been adapted as in the case of the dynamic time warping distance. The Jaccard coefficient is an alternative notion of similarity that "normalizes" similarity with the number of points considered: $J(X,Y) = |X \cap Y|/|X \cup Y|$ [32]. Hence, its value is in the interval of $[0,1]$ independent of the size of the compared sets. Furthermore, it can be easily converted into a distance by subtracting it from 1, which allows us to use a "standard" subsequent DBSCAN approach to search for clusters.

Interestingly, the basic concept of shared nearest neighbor similarity is analogous to the "sparsifying" approach used by Laplacians for robust spectral clustering [21]. Since computing the nearest neighbor similarity also "reduces" the noise in the original data, it might be interesting to compare the results of a subsequent DBSCAN with the results of a subsequent robust spectral clustering.

C. Quality indicators

1) Generalized Dunn index

There are a multitude of quality coefficients for clustering [33]. These parameters are particularly important for finding the optimal parameter settings for clustering methods when no *ground truths are* available. In the context of the work presented in this paper, there is no upfront knowledge regarding similarity groups of team moves. Clustering is explicitly used to find representatives of groups that will be used by experts to label the tactics used in application terms.

At least two aspects need to be considered when selecting a quality indicator for the extracted clustering model in the context of the presented work. Clustering methods that handle noise explicitly, such as DBSCAN, assign noise records to a separate noise cluster that must be excluded from the calculation of a quality indicator value. As a result, there are two extreme cases. In the first case, the parameter settings are chosen such that all the non-noise points are assigned to a single cluster or very few clusters. This is, for instance, the case when the search radius for similar points of DBSCAN is too large. The second extreme is the case when the search radius is rather small, such that most of the found clusters consist of only a single data point and are thus treated as noise. Consequently, there are only a few but well-separated clusters. The first case is indicated by a low number of noise points and clusters with a large distance between the contained points, while the latter case has only a relatively small number of non-noise points and a very small distance between the contained points.

Originally, Dunn introduced the idea of using the ratio of the *diameter* of a cluster to the distance to the closest neighboring cluster as a quality indicator for a clustering model [34]. This idea was later generalized by Bezdek and Pal, who introduced several notions of diameter (intra-cluster distance) and distance (inter-cluster distance) denoted as the Generalized Dunn Index (GDI) [35]. Since the chosen clustering approach does not compute any centroids, centroid-based variants have not been considered. To avoid oversensitivity to outliers, the average distances of the intra-cluster distances and the maximum distances of the inter-cluster distances were chosen as the underlying values to compute the GDI, which is also denoted as *GDI 2-2*. The Generalized Dunn Index is always positive, and the higher the value is, the better the clustering.

2) The side effect of excluding noise

The described clustering approach based on nearest neighbor similarity has 3 main parameters that can be varied to find an optimal clustering for a set of trajectories:

- The number of neighbors used to determine the similarity: The smaller the considered number of neighbors, the smaller the set of neighbors with a Jaccard coefficient greater than epsilon. The number of points treated as noise increases.
- The ε limits the ability to find "close" points: A small epsilon of DBSCAN results in small sets of close points that can be assigned to the same cluster. Consequently, the number of points treated as noise increases.
- The minimum number of points needed to form an initial cluster: A small number of close points results in many identified clusters. As a result, the number of points treated as noise increases because the cluster size decreases.

All three parameters have a direct impact on the clustering model and the quality indicators not only by resulting in differing numbers of clusters and sets of contained points but also because the number of points treated as noise is directly impacted. This is also reflected by the quality indicators. If a single parameter is varied from low to high, we obtain the same "behavior" for the quality indicators. Similarly, the silhouette coefficient and GDI 2-2 increase with an increasing parameter value, while the Davies–Bouldin index decreases with increasing parameter value (and the inverse Davies–Bouldin coefficient increases as well).

No indication of an optimal parameter setting can be derived from the course of the indicators' graphs. This is caused by the overlapping effect of a changing number of clusters and an increasing number of points considered noise; thus, these clusters are excluded from the quality indicator values. Hence, it is necessary to take the number of points considered valid into account as well as the number of clusters by weighting the clustering quality indicator. The basic concept of quality indicator weighting was introduced in [4].

A straightforward approach for weighting the quality indicator value is based on two simple observations (heuristics). Given that two clustering models have the same base quality indicator value, a clustering model consisting of more non-noise points is preferable because it represents more information of the input data. Second, if two clustering models have the same quality indicator values and the same number of non-noise points, then a model consisting of more clusters is considered preferable because it potentially allows for better differentiation of cases.

Simple weighting with the number of non-noise (or valid) points $N_v = |\{t_i^c\}|$ results in a weighted quality indicator whose value depends on the sample size. Thus, the relative number of non-noise points based on the size of the input sample $N=|\{t_i\}|$ is used rather than the absolute input size: $n_v = N_v/N$. Using the absolute number of identified trajectory clusters $G^C=|\{c_j\}|$ as an additional weight would overemphasize the importance of the number of clusters. Using the maximum number of clusters to normalize G^C would require knowing this number upfront. Thus, the ratio of the number of clusters to the sample size is used as the weight that represents the number of clusters: $g^C=|\{c_j\}|/N$.

A weighted clustering indicator value q_i^w can now be defined as the product of the original indicator value q_i and the two weights introduced in the previous section: $q_i^w = q_i\, n_v\, g^C$.

V. Applicability Study

A. Complexity and Runtimes

The concepts presented in this paper have been used in a real-world scenario of sports. In collaboration with a first league team and the first German Handball Bundesliga, *HBL*, the position data of all matches of the selected team in 2022, 2023, and first half of 2024 were collected as a basis for identifying the offensive tactics they played. The future objective is to be able to detect the played tactics of a team, which can then be used to automatically determine the performance of played tactics for teams and players.

The data consisted of 82 matches, from which a total of 12,366 team moves were extracted before a scoring attempt. A total of 3,020 moves of the 12,366 were offensive moves of the selected team, from which 23,674 trajectories were extracted. Since so-called fast break attacks are not of interest in the context of tactic recognition and some trajectories contain erroneous data, the data for the analysis were reduced to 2,089 team moves with 16,277 trajectories.

The practical evaluation was performed using MathWorks MATLAB™ R2024b version 24.2.0.2712019 running on Ubuntu 22.04.5 using a virtual machine with 16 vCPUs equipped with 32 GB of main memory.

The runtimes of the different computation steps of a complete single run with the "optimal" parameter settings (see Section V.B) are listed in TABLE I. The most time-consuming step of the data preparation is the calculation of the dynamic time warping distances for each pair of trajectories, which is of complexity n^2-n or $O(n^2)$ when n denotes the number of trajectories. The distance calculation itself is implemented using the classical dynamic programming approach with a complexity of mw with respect to m as the number of points contained in the trajectories and w as the window size, which results in an overall complexity of n^2mw (if w is defined as a ratio of m, then this results in $O(n^2m^2)$).

The computation of dynamic time warping distances is the most time-consuming step of a clustering run, even when searching for optimal parameter settings consisting of repeating subsequent steps. Thus, parallelizing the computation of the DTW distances helps to reduce the overall computation time significantly. Furthermore, it is advisable to use only the computed distances rather than the original trajectory data in the subsequent steps.

The second most time-consuming step after the computation of the dynamic time warping distances is the computation of the shared nearest neighbor similarity. Finding the k nearest neighbors ($k \ll n$) based on the previously computed distances is a set of nk simple searches of n-1 distances when n denotes the number of trajectories. However, computing a full matrix of nearest neighbors might be advantageous when variations in k need to be computed (see the following Section V.B). This is particularly needed when varying the optimal number of trajectories used to compute the shared neighbor similarity (see Section IV.C.2). The subsequent calculation of the Jaccard similarity coefficients to determine the value of the shared neighbor similarity is of complexity nk^2 and can be easily parallelized.

The runtime of the subsequent search for relevant frequent itemsets is negligible given the runtimes of the previous steps. However, this heavily depends on the number of relevant frequent items found.

TABLE I. Measured runtimes of computation steps

Computation Step	Runtime in seconds
Reading and filtering data	17
Computation of DTW distances	6684
Computation of similarities	926
Single DBSCAN clustering	24
Search for itemsets	3

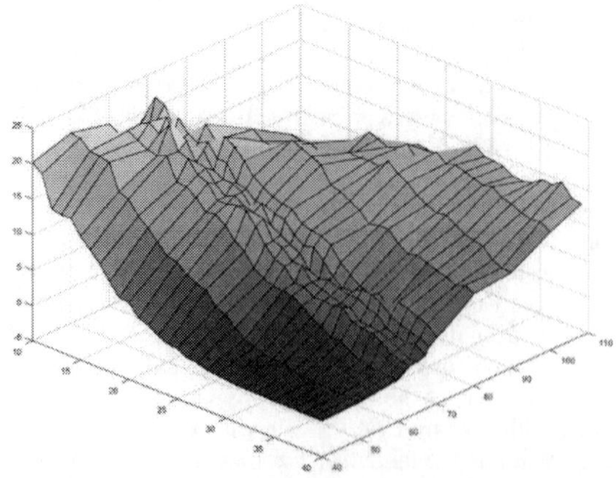

Figure 1. Weighted GDI 2-2 values with variations of the similarity window and the minimum number of points needed to be core

B. Optimal Parameters and Results

There are three parameters that can be varied while searching for trajectory clusters:

- The number of nearest neighbors that are checked to determine similar trajectories (*similarity window*).
- The minimum number of points that need to be close to be considered the core.
- The limit of the Jaccard coefficient that is used to determine "close" trajectories.

Rather than assuming that all of the mentioned parameters can be chosen arbitrarily, as indicated in [7], this paper follows a different approach. It is assumed that the ability of the Jaccard coefficient to identify close points is application dependent. In the given application context of trajectories of team handball players, we assume that at least 50% of the neighbors of two trajectories need to be "shared" in terms of the shared nearest neighbor approach to be considered "close trajectories". This translates into a minimum Jaccard coefficient of 0.40 (DBSCAN ε of 0.60).

Unfortunately, the similarity matrix directly depends on the number of nearest neighbors that are checked to determine similar trajectories. Hence, for each value of the similarity window, a similarity matrix needs to be computed, which is fairly time-consuming, as described in the previous Section V.A. However, to determine the optimal number of nodes in the similarity window, we compared 12 different cases.

The last parameter that was varied was the minimum number of similar trajectories for being core in terms of DBSCAN. This parameter was varied in the range of [10, 40] with the assumption that at least 10 close trajectories are needed to be considered core. Figure 1 depicts the weighted GDI 2-2 values when varying the similarity window and the limit for the minimum number of points to be considered core points in the sense of the DBSCAN algorithm. The colors indicate ranges of similar index values. Yellow is the color for the highest range, while blue is the color associated with the lowest range of index values. An interesting observation is that the diagonal direction is the same for the same levels of index values. It seems that a decreasing lower limit for core points can compensate for the effect of an increasing similarity window to some extent.

The global maximum of the weighted GDI 2-2 value is at a similarity window of 67 trajectories, and the minimum is 12 necessary points for a core. The value of the weighted GDI 2-2 peaks at 23.27 (based on a GDI 2-2 value of 1.05). The nearest neighbor similarity clustering identified 39 trajectory clusters that represented approximately 58% of the input trajectories. Approximately 42% of the trajectories are identified as noise. After coding the 1,757 team moves using trajectory cluster identifiers, 1,625 team moves with two or more non-noise trajectory cluster identifiers remained. A team move that contains fewer than two non-noise trajectory cluster identifiers is considered an individual move rather than a team tactical move.

The search for relevant frequent itemsets was performed using the 1,625 team moves of the previous step. The lower length limit was set to 3.0, which means that a trajectory is relevant only if the length of the trajectory is greater than 3.0 meters. This is an application-dependent limit and might differ between application scenarios. The absolute minimum support was set to 10, which means that an itemset is frequent if ten team moves support it. The search for frequent itemsets identified 143 itemsets of length 2, 41 itemsets of length 3, and 7 itemsets of length 4.

While 6,790 of the 16,277 trajectories were associated with a trajectory cluster, 1,562 team moves of 2,089 (approx. 75%) were associated with a team move group (or cluster of trajectory sets). 185 of 191 frequent itemsets were used to identify team move groups. The 6 "unused" frequent itemsets result from the criterion that was used to assign a frequent itemset with a team move (see Section III.B.4).

C. Application-level evaluation

To evaluate the results on an application level, the representative trajectories of each associated itemset were extracted. The contained trajectories were then visualized in a video presenting the tactical view and shown to team handball experts (coaches of the first league teams) to decide whether an actual team tactic was detected by the system and how the identified team tactic could be named. Figure 2 shows a snapshot of three example animations that have been presented to the coaches with their associated names. The seven offense players are depicted as green diamonds. Their trajectories are depicted as "sequences" of green diamonds. To depict the time aspect, the color of the diamonds starts with dark green and ends in bright green. Since the data have been transformed, such that attacks always occur from left to right, only the right half of the field is depicted.

The coaches were able to confirm that the extracted similarity groups actually represent team tactics rather than an arbitrary collection of team moves. The evaluation of the trajectory clusters was successful as well. The trajectory clusters clearly represented the different positions (roles) of the players whose trajectories were contained in a cluster. Some of the detected tactics were considered similar when

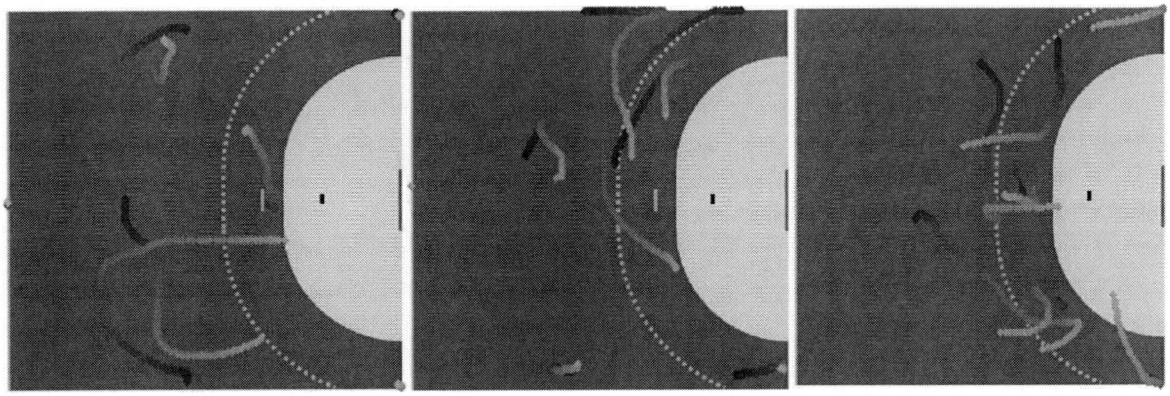

Figure 2. Three snapshots of animation videos of team move clusters: "empty crossing right", "runner from position 1", and "7 versus 6"

evaluated by human experts. A detailed analysis of the significant properties of these clusters needs to be performed to determine the key differences between the clusters. If the differences are correlated with the success or nonsuccess of the attacks, the analysis will help to identify the critical aspects of the realized tactics.

D. Development of a stable cluster model

Originally, 32 matches were used to evaluate the clustering approach, and it has been observed that the dataset was too small to extract a comprehensive set of trajectory clusters [7]. Thus, the approach is continuously evaluated with larger data sets. The results presented in this paper are based on the data of 82 matches. The originally identified trajectory clusters have been confirmed to a vast extent. However, some smaller clusters have been merged, which can be explained by the availability of additional data points that "bridge" the distance between the small clusters.

We assume that the set of identified trajectory clusters will eventually reach a stable state when enough trajectories can be used to extract the cluster model. Since we still observe significant changes in terms of clusters and the number of clustered trajectories, this stable state has not yet been reached. When a stable state has been reached, the number of team moves can be used as a "team move window size" to continuously extract models from the current data. These models can then be compared with previous models to detect significant changes in the set of clusters indicating significant changes in the applied tactics.

E. Using the model to automatically detect tactics

1) Concept for applying the model

The trajectory cluster model can be used to determine whether a detected trajectory belongs to one of the clusters based on the criteria that were used to extract the cluster model. Trajectories are assigned to a cluster if they belong to the core points of a cluster or if they are directly reachable from a core point in the sense of DBSCAN. Consequently, a trajectory is considered to belong to a previously identified trajectory cluster if it is directly reachable from any of the core points of that cluster.

To evaluate the criteria above, the similarity-based distances to all the core trajectories that are part of the clustering model need to be computed. If the distance to any of the core trajectories is less than the ε that was used to compute the clustering model, the trajectory belongs to the cluster of that core trajectory. If no core trajectory is within the ε distance, the trajectory that is checked cannot be assigned to any of the clusters and is treated as noise.

With the assignment of trajectories to clusters, team moves can be "recoded", as described in Section III.A.2) Then, it is determined whether the team move supports any of the previously identified items. In this case, the team move contains the tactic that is represented by the supported frequent itemset.

2) Performance aspects

Given the current set of data used to compute the clustering model, 3,912 core trajectories were identified. The SNN-based distance that is derived from the DTW distances needs to be computed for each trajectory contained in a team move to check the ε limit of direct reachability. A rough estimation using the measured time for calculating the distances and similarities of the trajectories to derive the cluster model is based on the total number of distances that have been calculated so far. In total, approximately 132 million distances and similarities were calculated, which took about 7,610 seconds. For the application case, approximately 27 thousand distances and similarities (3,912 times 7) need to be calculated, which can be estimated with an elapsed time of half a second. The encoding and the search for supported itemsets are in the millisecond range.

Overall, we can assume an upper limit of one second for automatically determining the tactics based on a stream of trajectories, which is fast enough for the given application case because the data of the team moves consist of 5.5 seconds of position data before an attempt happens, i.e., after an attempt, there is a minimum time gap of 5.5 seconds until we might have another set of trajectories of an attempt.

VI. Conclusion and Future Work

A clustering approach has been proposed to find similarity groups of team moves without the need for the upfront assignment of class labels. Using a two-step approach based on the concept of shared similarity and the dynamic time warping distance addresses multiple shortcomings of the original approach in finding similarity groups. In particular, the need for manual collection of data in addition to spatiotemporal data is avoided.

The results of the clustering of trajectories (the first step) can be verified by evaluating representatives of the identified clusters. From an application perspective, the representative trajectories should correspond with the intended individual moves of certain player types at the application level. With this application-level mapping, end-users are more likely to establish trust in the approach.

The second step of the search for similarity groups involves searching for relevant frequent itemsets that deviate from the usual approaches that try to solve the task via a clustering approach. Using the search for relevant frequent itemsets avoids the need for an explicit distance criterion, which is difficult to define and difficult to explain in the application context. Furthermore, the concept of relevance is important when combining individual trajectories with team tactical moves. It has been shown that the search for frequent itemsets can be efficiently combined with the concept of relevance of trajectories.

Rather than arbitrarily varying the parameters of the approach, application-level decisions have been made to limit the number of cases that need to be considered when looking for optimal parameters. With this approach, the quality of the trajectory cluster model and the number of "represented" trajectories increase significantly. Furthermore, the assignment ratio of team moves to team move similarity groups increased as well.

The results of the applicability study show that the approach works in a real-world scenario. Previously recognized problems due to the low amount of available data cannot be observed anymore. The long-term objective is to derive a stable model that allows us to assign a label to team moves during matches by using the previously extracted cluster model. Hence, we have a basis for a novel approach to take individual contributions to a team tactic into account when evaluating players' performance in the future.

ACKNOWLEDGMENT

The work presented in this paper was supported by the German Handball Federation (DHB) and the German Handball-Bundesliga GmbH (HBL). Furthermore, it was particularly supported by the first league handball teams, TVB Stuttgart and Frisch Auf! Göppingen.

REFERENCES

[1] Catapult, "Vector T7". [retrieved: January 2025]. Available: https://www.catapult.com/blog/vector-t7-white-paper

[2] Kinexon, "Perform LPS". [retrieved: January 2025]. Available: https://kinexon-sports.com/products/perform-lps/

[3] F. Schwenkreis, "An Approach to use Deep Learning to Automatically Recognize Team Tactics in Team Ball Games", in *Proceedings of the 7th Conference on Data Science, Technology and Applications.*, Porto: Scitepress, Jul. 2018, pp. 157–162.

[4] F. Schwenkreis, "Using the Silhouette Coefficient for Representative Search of Team Tactics in Noisy Data", in *Proceedings of the 11th Conference on Data Science, Technology and Applications*, Lisbon, Portugal: Scitepress, Jul. 2022, pp. 193–202.

[5] B. Aronov, S. Har-Peled, C. Knauer, Y. Wang, and C. Wenk, "Fréchet distance for curves, revisited", in *European symposium on algorithms*, Berlin, Heidelberg: Springer, 2006, pp. 52–63.

[6] P. J. Rousseeuw, "Silhouettes: A graphical aid to the interpretation and validation of cluster analysis", *Computational and Applied Mathematics*, no. 20, pp. 53–65, 1987.

[7] F. Schwenkreis, "Automated Detection of Trajectory Groups Based on SNN-Clustering and Relevant Frequent Itemsets", presented at the IEEE 10th International Conference on Data Science and Advanced Analytics (DSAA), Thessaloniki, Greece: IEEE, Oct. 2023, pp. 1–10.

[8] B. Kerner, Ed., "The Physics of Traffic", Springer, 2004.

[9] S. Wang, Z. Bao, J. S. Culpepper, T. Sellis, and X. Qin, "Fast large-scale trajectory clustering", *Proc. VLDB Endow.*, vol. 13, no. 1, pp. 29–42, Sep. 2019.

[10] Y. Liu et al., "ECMA: An Efficient Convoy Mining Algorithm for Moving Objects", in *Proceedings of the 30th ACM International Conference on Information & Knowledge Management*, Virtual Event Queensland Australia: ACM, Oct. 2021, pp. 1089–1098.

[11] G. Schrijvers, A. van Hoorn, and N. Huiskes, "The care pathway: concepts and theories: an introduction", *International journal of integrated care*, 2012. [retrieved: January 2025] Available https://doi.org/10.5334/ijic.812.

[12] V. Vogt, S. M. Scholz, and L. Sundmacher, "Applying sequence clustering techniques to explore practice-based ambulatory care pathways in insurance claims data", *European Journal of Public Health*, vol. 28, no. 2, pp. 214–219, Apr. 2018.

[13] U. Brefeld, J. Davis, J. V. Haaren, and A. Zimmermann, Eds., *Machine Learning and Data Mining for Sports Analytics*. Ghent, Belgium: Springer, 2020.

[14] U. Brefeld, J. Davis, J. Van Haaren, and A. Zimmermann, *Machine Learning and Data Mining for Sports Analytics*, Springer, 2021.

[15] U. Brefeld, J. Davis, J. Van Haaren, and A. Zimmermann, *Machine Learning and Data Mining for Sports Analytics*, Springer, 2022.

[16] P. Bauer and G. Anzer, "Data-driven detection of counterpressing in professional football: A supervised machine learning task based on synchronized positional and event data with expert-based feature extraction", *Data Min Knowl Disc*, vol. 35, no. 5, pp. 2009–2049, Sep. 2021.

[17] Y. Jiang and C. Bao, "Human-centered artificial intelligence-based ice hockey sports classification system with web 4.0", *Journal of Intelligent Systems*, vol. 31, pp. 1211–1228, 2022.

[18] S. Aghabozorgi, A. Seyed Shirkhorshidi, and T. Ying Wah, "Time-series clustering – A decade review", *Information Systems*, vol. 53, pp. 16–38, Oct. 2015.

[19] S. Aghabozorgi, T. Y. Wah, T. Herawan, and H. A. Jalab, "A Hybrid Algorithm for Clustering of Time Series Data Based on Affinity Search Technique", *The Scientific World Journal*, no. 3, 2014. [retrieved: January 2025] Available: https://doi.org/10.1155/2014/562194

[20] U. von Luxburg, "A tutorial on spectral clustering", *Statistics and Computing*, vol. 14, no. 4, pp. 395–416, 2007.

[21] A. Bojchevski, Y. Matkovic, and S. Günnemann, "Robust Spectral Clustering for Noisy Data: Modeling Sparse Corruptions Improves Latent Embeddings", in *Proceedings of the 23rd ACM SIGKDD International Conference on Knowledge Discovery and Data Mining*, Halifax NS Canada: ACM, Aug. 2017, pp. 737–746.

[22] M. Roux, "A Comparative Study of Divisive and Agglomerative Hierarchical Clustering Algorithms", *Journal of Classification*, no. 35, pp. 345–366, 2018, [retrieved: January 2025]. Available: https://doi.org/10.1007/s00357-018-9259-9.

[23] M. Ester, H.-P. Kriegel, J. Sander, and X. Xu, "A density-based algorithm for discovering clusters in large spatial databases with noise", in *Proceedings of the Second International Conference on Knowledge Discovery and Data Mining (KDD-96)*, Portland Oregon: AAAI Press, 1996, pp. 226–231.

[24] R. W. Hamming, "Error detecting and error correcting codes", *Bell System Technical Journal*, vol. 29, no. 2, pp. 147–160, 1950.

[25] R. Agrawal and R. Srikant, "Fast algorithms for mining association rules", in *Proceedings of 20th International Conference on Very Large Databases*, Santiago de Chile, Chile: Morgan Kaufmann, 1994, pp. 487–499.

[26] X. Zhao, S. C. Xinhui Zhang Pan Wang, and Z. Sun, "A weighted frequent itemset mining algorithm for intelligent decision in smart systems", *IEEE Access*, vol. 6, pp. 29271–29282, 2018.

[27] Y. Tao *et al.*, "A comparative analysis of trajectory similarity measures", *GIScience & Remote Sensing*, vol. 58, no. 5, pp. 643–669, Jul. 2021.

[28] R. Bellman and R. Kalaba, "On adaptive control processes", *IRE Transactions on Automatic Control*, vol. 4, no. 2, pp. 1–9, 1959.

[29] M. Brankovic et al., "(k, l)-Medians Clustering of Trajectories Using Continuous Dynamic Time Warping", in *Proceedings of the 28th International Conference on Advances in Geographic Information Systems*, Nov. 2020, pp. 99–110.

[30] J. Paparrizos, C. Liu, A. J. Elmore, and M. J. Franklin, "Debunking Four Long-Standing Misconceptions of Time-Series Distance Measures", in *Proceedings of the 2020 ACM SIGMOD International Conference on Management of Data*, Portland OR USA: ACM, Jun. 2020, pp. 1887–1905.

[31] L. Ertöz, M. Steinbach, and V. Kumar, "Finding Clusters of Different Sizes, Shapes, and Densities in Noisy, High Dimensional Data", in *Proceedings of the 2003 SIAM International Conference on Data Mining*, Society for Industrial and Applied Mathematics, May 2003, pp. 47–58.

[32] P. Jaccard, "The Distribution of the Flora in the Alpine Zone", *New Phytologist*, vol. 11, no. 2, pp. 37–50, 1912.

[33] Bernard Desgraupes, "Clustering Indices." [retrieved: January 2025] Available: https://de.scribd.com/document/268911122/Cluster-Criteria

[34] J. C. Dunn, "Fuzzy Relative of the ISODATA Process and Its Use in Detecting Compact Well-Separated Clusters", *Journal on Cybernetics*, vol. 3, no. 3, pp. 32–57, 1973.

[35] J. C. Bezdek and N. R. Pal, "Some new indexes of cluster validity", *IEEE Trans. Syst., Man, Cybern. B*, vol. 28, no. 3, pp. 301–315, Jun. 1998.

On the Performance of Query Optimization Without Cost Functions and Very Simple Cardinality Estimation

Daniel Flachs
University of Mannheim
Mannheim, Germany
email: flachs@uni-mannheim.de

Guido Moerkotte
University of Mannheim
Mannheim, Germany
email: moerkotte@uni-mannheim.de

Abstract—**In order to enable a fast time to market for new Database Management Systems (DBMS), we introduce two simple, very easy to implement cardinality estimators and a build-plan method that does not require any cost function. Experimentally, we demonstrate that different plan generators incorporating these ideas are quite competitive on the Join Order Benchmark (JOB): the join ordering algorithm DPccp yields plans that are at most a factor of 2.10 away from the optimum without using any runtime-dependent cost function if cardinalities are known. Thus, using our approach obviates the effort of implementing sophisticated cardinality estimation methods and cost functions in a first version of a DBMS.**

Keywords-query optimization; hash join; cardinality estimation; cost function; plan generation.

I. INTRODUCTION

Developing a new DBMS like DuckDB [1] from scratch is a challenging task. To achieve a short time to market, compromises in different modules of the system are required. Two modules that are tedious to implement and test in the context of query optimization are cardinality estimation (CE) and cost functions (CF).

We observed recent attempts in the literature to simplify query optimization [2][3]. However, these approaches are often monolithic in the sense that they touch many of the different 'moving parts' involved in query optimization, without proper modularization. Those modules are

- the **join ordering algorithm** that decides the logical join order,
- the **build-plan procedure** (BP) used for selecting the most suitable physical (join) operator,
- the **cardinality estimator** (CE) that provides a cardinality estimate for any subset of relations considered by the plan generator,
- the **cost function** (CF) to compare the cost of two plan alternatives.

If implemented correctly, the modules are independent of each other and can be exchanged without changing the rest of the system, allowing for an extensive evaluation of the different combinations.

In this paper, we pose the questions: How simple can cardinality estimation, cost function, and build plan procedure become while still yielding query evaluation plans (QEPs) with an acceptable quality? Can we get acceptable plans even if BP does not use any cost function at all?

In order to answer these questions, we propose two very simple cardinality estimators: CE_{base} and CE_{sel}. Further, we propose a new build-plan method for query optimization (BP_{smart}) that does not require any cost function and instead relies only on cardinality estimates for its decisions. This is in stark contrast to the traditional build-plan procedure BP_{trad}: Given two (sub-) plans to be joined, it decides on the argument order (e.g., build vs. probe side for hash joins) as well as on the actual join implementation to be used in a purely cost-based manner.

In order to evaluate and compare the performance of the newly proposed cardinality estimators and build-plan method, we implemented the Plan Generator Benchmarking Framework *PgBench*. It allows to orthogonally test the performance of combinations of join ordering algorithms, cardinality estimation methods, cost functions, and build-plan procedures. In this paper's evaluation, we report on the performance of the join ordering algorithms DPccp [4], GooCost [5], and GooCard [6][7]. As cardinality estimation methods, we use our new cardinality estimators CE_{base} and CE_{sel} as well as the existing $CE_{IA\text{-}M}$, which relies on the independence assumption, and CE_{tru}, which provides the true cardinalities. As cost functions, we use CF_{tru} and CF_{est} for the true and estimated execution costs. The build-plan procedures used are BP_{trad} and the newly designed BP_{smart}. As the set of queries, we use the Join Order Benchmark (JOB) [8].

Since the implementations of BP_{smart} and CE_{base} or CE_{sel} require only little effort while being quite competitive, as seen in the evaluation, this will help to achieve a short time to market for a new DBMS.

The rest of the paper is organized as follows. Section II presents basic notions like plan class, ccp, uniqueness, and loss factor of a plan. Section III gives the details on the new cardinality estimation methods CE_{base} and CE_{sel} as well as for $CE_{IA\text{-}M}$. Section IV introduces the different join implementations as well as the derivation of the cost functions CF_{tru} and CF_{est} for them. Sections V and VI introduce the two build-plan procedures BP_{trad} and BP_{smart}. Section VII contains the evaluation. An overview of related work is given in Section VIII. Section IX concludes the paper.

II. PRELIMINARIES

For a given conjunctive query, we denote by $\mathcal{R} := \{R_1, \ldots, R_n\}$ the set of relations in its `from`-clause. The set of attributes of R_i is denoted by $\mathcal{A}(R_i)$. The single-table selection predicates for R_i are denoted by p_i. Equijoin predicates for a join between R_i and R_j are denoted by $p_{i,j}$. The attributes accessed by a selection or join predicate p are denoted by $\mathcal{F}(p)$. The *query graph (QG)* has the relations in $\mathcal{R}$ as nodes and an edge connecting R_i and R_j for every join predicate $p_{i,j}$. A *plan class* $S \subseteq \mathcal{R}$ is a subset of relations inducing a connected subgraph of the query graph. In this case, we write $\text{pc}(S)$. The set of attributes of a plan class $\text{pc}(S)$ is defined as $\mathcal{A}(S) := \bigcup_{R_i \in S} \mathcal{A}(R_i)$. Two subsets $S_1, S_2 \subseteq \mathcal{R}$ form a *ccp* if $\text{pc}(S_1)$, $\text{pc}(S_2)$, $S_1 \cap S_2 = \emptyset$, and there is at least one join predicate $p_{i,j}$ such that $R_i \in S_1$ and $R_j \in S_2$ [4]. In this case, we write $\text{ccp}(S_1, S_2)$. The abbreviation ccp stands for a *csg-cmp-pair*, which refers to a pair of connected subgraphs *(csg)* of the query graph that are complements *(cmp)* to each other. The notion of a ccp was first introduced by Moerkotte and Neumann [4]. To find the optimal join order using dynamic programming, all such pairs must be enumerated.

For both our new build-plan procedure and our new cardinality estimators, we need to determine whether the join predicate connecting the two sets of a $\text{ccp}(S_1, S_2)$ implies uniqueness on S_1 and/or S_2.

The join predicate $P(S_1, S_2)$ for $\text{ccp}(S_1, S_2)$ is the conjunction of all $p_{i,j}$ such that $R_i \in S_1$ and $R_j \in S_2$. The set of join attributes of S_i ($i = 1, 2$) is then defined as $\mathtt{J}(S_i, (S_1, S_2)) := \mathcal{F}(P(S_1, S_2)) \cap \mathcal{A}(S_i)$. We say that $\text{ccp}(S_1, S_2)$ determines S_i uniquely if $\mathtt{J}(S_i, (S_1, S_2))$ is a (super-) key of S_i. If $\mathtt{K}(S_i)$ denotes the set of keys of S_i, we can express this as

$$\mathfrak{U}(S_i, (S_1, S_2)) := \exists \kappa \in \mathtt{K}(S_i) : \kappa \subseteq \mathtt{J}(S_i, (S_1, S_2)).$$

In order to derive this uniqueness, we need to determine the keys for plan classes S. We start by assuming that for every relation R_i the set of primary and secondary keys is given as $\mathtt{K}(\{R_i\})$. For a $\text{ccp}(S_1, S_2)$, we can determine the set of keys $\mathtt{K}(S_1 \cup S_2)$ as follows. If there exists a key $\kappa \in \mathtt{K}(S_i)$ such that $\kappa \subseteq \mathtt{J}(S_i, (S_1, S_2))$, then $\mathtt{K}(S_{3-i}) \subseteq \mathtt{K}(S_1 \cup S_2)$. Note that S_{3-i} refers to S_2 for $i = 1$ and vice versa. If this is true for neither S_1 nor S_2, then $\mathtt{K}(S_1 \cup S_2)$ contains $\kappa_1 \cup \kappa_2$ for all $\kappa_1 \in \mathtt{K}(S_1)$ and for all $\kappa_2 \in \mathtt{K}(S_2)$.

To measure the error between a true value and an estimate, we use the *q-error* [9]. Let $x > 0$ be some true value and $\hat{x} > 0$ be its estimate, then $\text{qerr}(x, \hat{x}) := \max\left(\frac{x}{\hat{x}}, \frac{\hat{x}}{x}\right)$.

Finally, we define the *loss factor* of a plan. Given a plan class S and a plan P for S, we define the *loss factor* of P as the true cost of P divided by the true cost of the overall best plan for S. This is a value greater or equal to 1.

III. Cardinality Estimation

In this section, we present two new cardinality estimators (CE_{base} and CE_{sel}) which only differ in their treatment of single-table selection predicates. CE_{base} ignores them, whereas CE_{sel} takes them into account. We start by defining both estimators for plan classes containing a single relation $R_i \in \mathcal{R}$:

$$\text{CE}_{\text{X}}(\{R_i\}) := \begin{cases} |R_i| & \text{if X = 'base'} \\ |\sigma_{p_i}(R_i)| & \text{if X = 'sel'} \end{cases} \tag{1}$$

Then, for general plan classes, we define

$$\text{CE}_{\text{X}}(S) := \min_{\text{ccp}(S_1, S_2):\ S = S_1 \cup S_2} \text{CE}_{\text{X}}(S, (S_1, S_2)) \tag{2}$$

where, with $\mathfrak{U}(S_i)$ abbreviating $\mathfrak{U}(S_i, (S_1, S_2))$,

$$\text{CE}_{\text{X}}(S, (S_1, S_2)) := \begin{cases} \min(\text{CE}_{\text{X}}(S_1), \text{CE}_{\text{X}}(S_2)) & \text{if } \mathfrak{U}(S_1) \land \mathfrak{U}(S_2) \\ \text{CE}_{\text{X}}(S_1) & \text{if } \neg\mathfrak{U}(S_1) \land \mathfrak{U}(S_2) \\ \text{CE}_{\text{X}}(S_2) & \text{if } \mathfrak{U}(S_1) \land \neg\mathfrak{U}(S_2) \\ \text{CE}_{\text{X}}(S_1) \cdot \text{CE}_{\text{X}}(S_2) & \text{if } \neg\mathfrak{U}(S_1) \land \neg\mathfrak{U}(S_2) \end{cases}$$

The idea is to use the smaller cardinality if both arguments are unique, the cardinality of the cross product if neither argument is unique, and the cardinality of the non-unique side if one of the argument relations is unique but not the other. Clearly, both cardinality estimators may produce overestimates and never produce underestimates, given true inputs. Note that for an implementation of CE_{sel}, an estimation procedure for $|\sigma_{p_i}(R_i)|$ is required. We propose to use sampling, as it is easy to implement and universally applicable [10]–[12].

For comparison in our evaluation, we also use the cardinality estimator $\text{CE}_{\text{IA-M}}$, which applies the independence assumption using the multiplicative rule [13]:

$$\text{CE}_{\text{IA-M}}(S) := \prod_{R_i \in S} |\sigma_{p_i}(R_i)| \cdot \prod_{p_{i,j}:\ R_i, R_j \in S} \text{sel}(p_{i,j}) \tag{3}$$

where $\text{sel}(p_{i,j}) := \frac{|R_i \bowtie_{p_{i,j}} R_j|}{|R_i| \cdot |R_j|}$ is the (true) selectivity of $p_{i,j}$.

IV. Cost Functions

This section first introduces the two physical hash join operators and their variants before outlining how their cost functions are derived from runtime experiments.

A. The Join Implementations

We consider two physical main-memory hash join operators: the *chaining hash join (CH-join)* and the *3D hash join (3D-join)* [14]. Their main difference is the hash table data structure which is built and probed during the join. The CH-join uses a hash table that resolves collisions by collecting all colliding keys into one linked list for each hash table bucket. The 3D-join uses a 3D hash table that groups duplicate keys together in a hierarchical collision chain organization with main and sub nodes. Further, for both physical operators, we consider two variants for the physical design of collision chain nodes, and three prefetching variants.

For the **collision chain node design**, there is an *unpacked (upk)* and a *packed (pkd)* variant. The unpacked variant is the original implementation [14], where each (main, sub) collision chain node stores one tuple pointer. The idea behind the packed variant is to improve the cache line utilization of a single collision chain node. Here, each collision chain node of the CH-join can store three tuple pointers. For the 3D-join, each main node stores five tuple pointers with equal join attribute values, and each sub node stores three tuple pointers.

Prefetching is a known technique to hide memory latencies of cache misses [15][16]. We therefore augment the non-prefetching implementations *(NoPF)* of the physical join operators by two prefetching approaches: *rolling prefetching (RoPF)* and *asynchronous memory access chaining (AMAC)* [16]. AMAC maintains a small ring buffer that keeps track of the processing state of tuples during the build or probe phase of a hash join, where the number of states depends on the join phase and physical hash table implementation. Before the next processing step of buffer element i, e.g., accessing a hash table bucket or inserting into a hash table collision chain node, a prefetch is issued for the necessary memory address, and processing continues with the next buffer element. Only after all other ring buffer elements have been examined, processing continues for buffer element i, giving the prefetch issued by i time to be completed. In AMAC, both hash table directory entries and hash table collision chain nodes are prefetched. There is obviously a tradeoff between the time saved due to hidden latencies, and the time lost due to branch misprediction penalties for handling the different AMAC states. As a compromise between NoPF and AMAC, we also implemented RoPF, which only prefetches the hash table directory entries, but not the collision chain nodes, simplifying the prefetching logic.

In summary, for each join in the plan, we can choose between any of the following 36 physical operators and implementations:

$$\{\text{CH-join}, \text{3D-join}\} \times \{\text{upk}, \text{pkd}\} \times \{\text{NoPF}, \text{RoPF}, \text{AMAC}\}^2$$

Note that the prefetching variants can be applied independently to the build and probe phase of a single hash join (hence the squared term), while both phases must agree on the physical node design.

B. Derivation of Cost Functions

The cost functions in this paper are constructed from measurements of runtime experiments. We therefore first briefly outline the experimental setup to obtain the measurements before describing the process of constructing cost functions from them.

1) Runtime Experiments: We measure the runtime of a single key/foreign key join between a key relation R and a foreign key relation S separately for the build and probe phase. The cardinalities for R and S are varied between 2^0 and 2^{30} in half-steps of powers of two, i.e., $\{t \cdot 2^i \mid i \in [0, 30], t \in \{1, 1.5\}\}$. For validation purposes, we additionally measured runtimes for $t \in \{1.3, 1.7\}$ that were not used for cost function generation. To generate the foreign keys for S, we draw $|S|$ random samples according to a (1) uniform and (2) standard Zipf distribution from a domain $[0, |S|/2^d - 1]$ with $d \in [0, 10]$ for build, and $d \in [0, 6]$ for probe. This results in a total of 43 489 input parameter combinations (without validation). For each of these combinations, we evaluate $|\{\text{CH}, \text{3D}\} \times \{\text{upk}, \text{pkd}\} \times \{\text{NoPF}, \text{RoPF}, \text{AMAC}\}| = 12$ build and the same number of probe runs. The runtimes are recorded in terms of timestamp counter clock ticks, a proxy for the wall-clock time [17, Chap. 18.17].

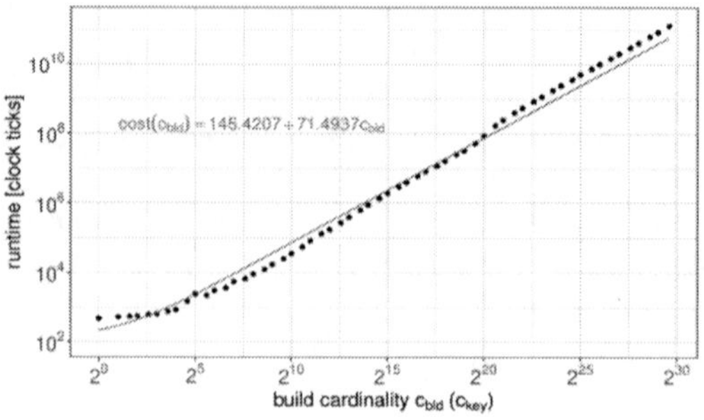

(a) unique build.

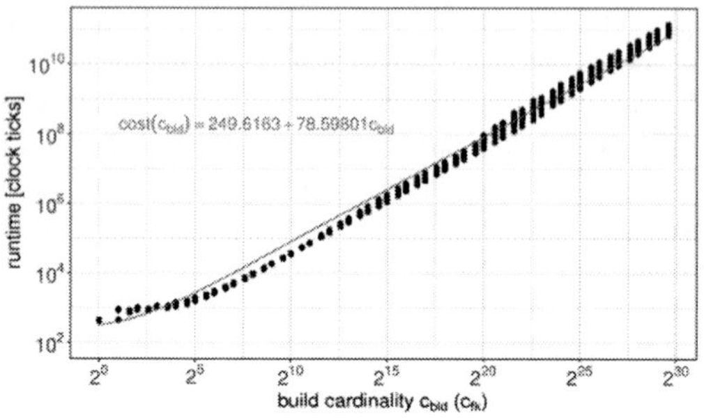

(b) non-unique build.

Figure 1. Measured runtimes and approximated functional cost functions for the build phase of CH-upk-NoPF.

2) Cost Function Generation: To turn the discrete data points from our runtime experiments into continuous cost functions usable by the query optimizer, we apply existing methods [9][18] to find an approximation function from a set of functions (e.g., constants, linear functions, or polynomials of a certain degree) that minimizes the maximum q-error between the true measured value and the values given by the function.

For each of the physical operator implementation variants from Section IV-A, we create separate cost functions for each case from $\{\text{build}, \text{probe}\} \times \{\text{unique}, \text{non-unique}\}$, i.e., for the two join phases and depending on the uniqueness of the build side. Each of the cost functions uses a subset of the following input parameters: the join's input build, probe and output cardinality (c_{bld}, c_{prb}, c_{res}), and the number of distinct values of the foreign key attribute ($nodv$).

Further, we construct cost functions in two levels of precision: a more precise, but also more complex *tabulated* cost function, and a less precise, but less complex *functional* cost function.

a) Functional Cost Functions: The functional cost functions capture all measurements of a single case (build/probe, uniqueness) in a single mathematical function. All cost functions are linear. For the build phase, the cost functions are 1-dimensional with c_{bld} as input, while they are 3-dimensional for the probe phase and use all three join cardinalities, c_{bld}, c_{prb}, and c_{res}. Figure 1 shows the measured runtimes from our experiments for the CH-join (unpacked, no prefetching) alongside the respective cost functions generated by approximation. Observe that the cost functions produce both over- and underestimates.

b) Tabulated Cost Functions: The tabulated cost functions use one- or two-dimensional lookup tables (hence the name) to map a subset of the measurements to an n-dimensional approximation function for that subset. To compute cost values for input parameter values between lookup table entries, we apply (bi-) linear interpolation. If the requested values are above the maximum values in the table, linear extrapolation is applied.

First, consider the build phase. For unique builds, each lookup table entry simply maps c_{bld} to the respective experimental runtime (a constant). For non-unique builds, each c_{bld} is associated with either a constant or a linear function that takes $nodv$ as its only input.

All lookup tables for the probe phase are two-dimensional in c_{bld} and c_{prb}. In the case of unique build sides, each lookup table entry contains a constant or a linear function that takes $nodv$ as its only input. For non-unique builds, each $(c_{\text{bld}}, c_{\text{prb}})$-pair is associated with either a constant or a two-dimensional linear function in $nodv$ and c_{res}.

We use the more precise tabulated cost functions as CF_{tru} and the simpler but less precise functional cost functions as CF_{est}. Note that both CF_{tru} and CF_{est} exhibit errors with regard to the true execution cost of the respective join, even though the name CF_{tru} might suggest otherwise. We merely assume CF_{tru} to be the true execution cost in order to have a notion of 'true optimality'. This distinction, however, is of minor importance towards the evaluation in Section VII, as the maximum q-error of CF_{tru} is well below 2 across all runtime measurements.

V. Traditional Build Plan

The task of finding a join tree for a given query graph can be split into two distinct problems: the *join ordering* that decides which relations and subtrees to join next, and the *operator selection* that chooses the most suitable physical operator and argument order to join two subtrees. The former problem is tackled by optimal join order ordering algorithms like DPccp [4], or heuristics, like GooCard [6][7] and GooCost [5].

Algorithm 1 DPccp [4].

```
1:  function DPCCP
2:     Input: a connected QG w/ relations R = {R_1, ..., R_n}
3:     Output: an optimal bushy join tree
4:     for all R_i ∈ R do BestPlan({R_i}) ← R_i
5:     for all ccp(S_1, S_2), S ← S_1 ∪ S_2 do
6:        T_1 ← BestPlan(S_1), T_2 ← BestPlan(S_2)
7:        T_curr ← BUILDPLAN(T_1, T_2)
8:        if COST(BestPlan(S)) > COST(T_curr) then
9:           BestPlan(S) ← T_curr
10:    return BestPlan(R)
```

For illustration, we show DPccp in Algorithm 1: it iterates over each ccp and stores the (cost-wise) optimal join tree for each plan class in a data structure. The operator selection problem is decided by a build-plan subroutine (called in Line 7), like BP_{trad}, shown in Algorithm 2, or BP_{smart} (see Section VI). Build-plan decides which physical join operators (*Implementations* in Algorithm 2, Line 5) are considered, and which argument order is better, $T_1 \bowtie T_2$ or $T_2 \bowtie T_1$. In case of BP_{trad}, this requires a total of 72 cost function evaluations for both join argument orders and 36 physical operator variants (see Section IV-A).

Algorithm 2 BP_{trad} [19, p. 62].

```
1:  function BUILDPLANTRAD(T_1, T_2)
2:     Input: two join trees T_1, T_2
3:     Output: the best join tree for joining T_1 and T_2
4:     BestTree ← null, COST(BestTree) ← ∞
5:     for each impl ∈ Implementations do
6:        T ← T_1 ⋈^impl T_2
7:        if COST(BestTree) > COST(T) then
8:           BestTree ← T
9:        T ← T_2 ⋈^impl T_1
10:       if COST(BestTree) > COST(T) then
11:          BestTree ← T
12:    return BestTree
```

TABLE I. Maximum loss factors of two plans for different input cardinality situations.

	CH-upk-(RoPF, RoPF)-bun	3D-upk-(RoPF, AMAC)-bnu
$2\|R\| < \|S\|$	1.79	7.71
$2\|R\| = \|S\|$	1.54	2.63
$\|R\| = \|S\|$	1.40	2.84
$\|R\| = 2\|S\|$	1.43	2.55
$\|R\| > 2\|S\|$	4.45	2.01

Both join ordering and build-plan apply cost functions to decide on the best alternative. Note that both could use different cost functions independently of each other: Build-plan could use cost functions based on join runtime (Section IV-B), whereas DPccp could pick the partial plan based on CF_{cout} (minimizing the sum of the cardinalities of the intermediate results) [20]. This flexibility is exploited for BP_{smart} in our evaluation.

VI. Smart Build Plan

It is the main goal of the build-plan procedure BP_{smart} to make all required decisions based only on the cardinalities of the input relations to the join, i.e., without any reference to a cost function. At the same time, it should try to minimize the loss factor of the produced (partial) plan.

One common heuristics is to always build on the smaller input relation. However, we will see that we need to slightly relax this rule when determining the order of the join's arguments.

Having chosen the build relation, we need to decide which join implementation and variant to apply. Based on runtime experiments [14], we come up with the following rule: If the join attributes of the build side cover a key of the build side (i.e., unique builds, *bun* for short), the CH-join is the clear favorite. Otherwise, for non-unique builds (*bnu* for short), the 3D-join is used. Since the experiments indicate that the packed versions of the algorithms only provide limited improvements in rare cases (many duplicates, uniform distribution), BP_{smart} uses only the unpacked versions.

For the CH-join with unique build sides, rolling prefetching is superior to AMAC in most cases. Further, it provides only little overhead for small builds compared to no prefetching. This applies to both the build and the probe phase. For the 3D-join with non-unique build-sides, we found that rolling prefetching is the best compromise for build and AMAC the best compromise for probe.

Let us come back to the problem of deciding which relation to use for the build. If we have a key relation R and a foreign key relation S, we can use the CH-join with build on R and the 3D-join with build on S. For a key/foreign key join, Table I contains the maximum loss factor of these two plans for different cardinality situations of the input relations. The header line shows the two physical plans used. For instance, 3D-upk-(RoPF, AMAC)-bnu refers to the 3D-join in the unpacked variant with a non-unique build side using rolling prefetching for build and AMAC for probe. The table indicates that we should use the CH-join if the cardinality of the key relation does not exceed twice the cardinality of the foreign key relation. If both or none of the input relations are unique, BP_{smart} uses the smaller one as the build relation.

The details of BP_{smart} are given in Algorithm 3. By convention, the right-hand side of the join symbol is the build side. The function *relset* returns the set of relations joined in a (partial) plan. Further, we abbreviate CH-upk-(RoPF, RoPF) by ch and 3D-upk-(RoPF, AMAC) by $3d$. We always use a CH-join if the build is unique, otherwise we use the 3D-join. If neither or both of the input relations are unique, then the build is on the smaller input. In Line 16, we make sure that T_1 is unique and T_2 is non-unique by a conditional swap. Then, we proceed as deduced in the above analysis.

Algorithm 3 BP_{smart}.

1: **function** BUILDPLANSMART(T_1, T_2)
2: **Input:** two join trees T_1, T_2
3: **Output:** a join tree for joining T_1 and T_2
4: **if** $\mathfrak{U}(relset(T_1)) \wedge \mathfrak{U}(relset(T_2))$ **then**
5: **if** $card(T_1) \leq card(T_2)$ **then**
6: $ResultTree \leftarrow T_2 \bowtie^{ch} T_1$
7: **else**
8: $ResultTree \leftarrow T_1 \bowtie^{ch} T_2$
9: **return** $ResultTree$
10: **if** $\neg\mathfrak{U}(relset(T_1)) \wedge \neg\mathfrak{U}(relset(T_2))$ **then**
11: **if** $card(T_1) \leq card(T_2)$ **then**
12: $ResultTree \leftarrow T_2 \bowtie^{3d} T_1$
13: **else**
14: $ResultTree \leftarrow T_1 \bowtie^{3d} T_2$
15: **return** $ResultTree$
16: **if** $\mathfrak{U}(relset(T_2))$ **then** $swap(T_1, T_2)$
17: **if** $card(T_1) \leq 2 \cdot card(T_2)$ **then**
18: $ResultTree \leftarrow T_2 \bowtie^{ch} T_1$
19: **else**
20: $ResultTree \leftarrow T_1 \bowtie^{3d} T_2$
21: **return** $ResultTree$

VII. EVALUATION

As our evaluation dataset and workload, we use the Join Order Benchmark (JOB) [8][21]. It consists of 33 query templates on the Internet Movie Database (IMDb) schema and dataset, which are instantiated as 113 queries, each with four to 17 relations. Queries from the same template only differ in their conjunctively connected single-table filter predicates. JOB's challenging analytical select-project-join queries justify its frequent use in the literature to evaluate the quality of plan generators [2][3][22]–[24].

In order to answer the question from the introduction whether the generation of acceptable plans is possible without a runtime-related cost function, we consider two cases in particular. Recall that our BP_{smart} makes all decisions without any notion of cost. We consider the combination of BP_{smart} with DPccp and CF_{cout}, and with GooCard, entirely eliminating runtime-related cost functions from the process of plan generation. We compare this to the usual approach where both join ordering and build-plan rely on runtime-related cost functions like CF_{tru} and CF_{est}.

Before going into a more detailed analysis, let us illustrate the general impact of join ordering on the given workload: If we modify plan generation such that it produces the overall

TABLE II. LOSS FACTORS FOR DPCCP, GOOCARD, AND GOOCOST.

DPccp					
BP	CF	CE_{tru}	$CE_{IA\text{-}M}$	CE_{base}	CE_{sel}
trad	tru	1.00	10.51	16.47	7.39
		1.00	1.39	2.32	1.98
	est	3.99	3.02	6.90	6.18
		1.82	1.51	2.67	2.57
smart	tru	2.52	17.88	6.84	6.09
		1.43	1.75	2.45	2.38
	est	2.10	20.98	6.94	6.09
		1.38	1.65	2.41	2.30
smart	cout	2.10	18.11	6.10	6.10
		1.41	2.07	2.57	2.31
GooCost					
BP	CF	CE_{tru}	$CE_{IA\text{-}M}$	CE_{base}	CE_{sel}
trad	tru	2.18	2.51	7.39	8.80
		1.11	1.32	2.10	2.19
	est	4.54	3.66	8.92	8.20
		1.97	1.72	2.57	2.52
smart	tru	2.86	2.32	6.90	12.23
		1.57	1.57	2.17	2.39
	est	2.66	2.66	6.90	5.74
		1.45	1.49	2.14	2.17
GooCard					
BP	CF	CE_{tru}	$CE_{IA\text{-}M}$	CE_{base}	CE_{sel}
trad	tru	1.43	13.52	7.39	13.65
		1.06	1.51	2.16	2.35
	est	3.15	6.52	8.92	8.30
		1.73	1.78	2.59	2.61
smart	—	2.10	15.53	6.90	6.71
		1.42	1.94	2.27	2.32

worst possible join order (using DPccp with cost maximization), the maximum (average) loss factor across all JOB queries is 35 785 (1168). As we will see, all subsequent loss factors are orders of magnitude away from this worst case.

Table II contains the plan loss factors for DPccp, GooCost, and GooCard. The first two columns indicate which combination of build-plan (BP) and cost function (CF) was applied. Typically, BP_{trad} uses the same cost function as the join ordering algorithm. In contrast, BP_{smart} does not require a cost function, so CF refers *only* to the join ordering algorithm for these cases. For every BP-CF-combination, there exist two lines. The first (second) line contains the maximum (average) loss factor taken over all JOB queries. The four numbers in each row correspond to the cardinality estimator used, as shown in the header line.

To start our discussion, we consider the first two lines of Table II. Here, we evaluate the loss factor of DPccp under BP_{trad} and CF_{tru} for different cardinality estimators. We see that the maximum loss factor under CE_{tru} is 1, which is the smallest possible value. The maximum (average) loss factors for the different cardinality estimators are 10.51 (1.39) for $CE_{IA\text{-}M}$, 16.47 (2.32) for CE_{base}, and 7.39 (1.98) for CE_{sel}. Thus, CE_{base} is the worst for the maximum loss factor and CE_{sel} the best. For the average, $CE_{IA\text{-}M}$ is the best. However, we still use CF_{tru} here, thus no cost function errors occur.

The next two lines are for BP_{trad} and CF_{est}. Here, the more realistic case of an erroneous cost function is evaluated. We see that even for the true cardinalities CE_{tru}, the maximum loss factor is 3.99 and the average is 1.82. Interestingly, the errors of $CE_{IA\text{-}M}$ and CF_{est} seem to compensate each other in the worst case, since the maximum loss factor decreases to 3.02, whereas the average increases to 1.51. For our new cardinality estimators, we have maximum loss factors of 6.90 (CE_{base}) and 6.18 (CE_{sel}), and average loss factors of 2.67 (CE_{base}) and 2.57 (CE_{sel}). Thus, they perform worse than $CE_{IA\text{-}M}$ if BP_{trad} and CF_{est} are in use.

This picture changes if we consider our newly introduced build-plan procedure BP_{smart}, where DPccp uses CF_{est}. Here, the maximum loss factor of $CE_{IA\text{-}M}$ (20.98) is far worse than that of CE_{base} (6.94) and CE_{sel} (6.09). On average, however, $CE_{IA\text{-}M}$ performs slightly better than these.

One of the goals of this paper is to provide a possibility to generate plans without the need for any runtime-related cost functions as, e.g., constructed in Section IV. For that purpose, we evaluated DPccp using CF_{cout} [20], which sums up the intermediate result sizes of joins as provided by the cardinality estimator in place. Further, BP_{smart} makes all decisions based only on cardinalities and uniqueness properties. Thus, we next report on the performance of DPccp without any reference to a runtime-related cost function. The last two lines of the DPccp-block contain the loss factors for this scenario. We see that if true cardinalities are used, the maximum loss factor is only 2.10 with an average of 1.41. If $CE_{IA\text{-}M}$ is used, the maximum loss factor increases to 18.11, which is much higher than the worst case for CE_{base} and CE_{sel}. On average, $CE_{IA\text{-}M}$ performs slightly better than CE_{base} and CE_{sel}. Further, under these conditions, both CE_{base} and CE_{sel} perform slightly better than under BP_{trad} and CF_{est}. Comparing CE_{base} and CE_{sel}, we see that the maximum loss factor is the same, but CE_{sel} performs slightly better than CE_{base} on average. This indicates that in a first version of a new DBMS, one could use CE_{base}. In some later version, CE_{sel} could be implemented. Remember that CE_{sel} requires the estimation of single-table selection predicates, for which, e.g., sampling needs to be implemented.

Let us now turn to the heuristics GooCost and GooCard. Under optimal conditions (BP_{trad}, CF_{tru}, CE_{tru}), the maximum (average) loss factors of GooCost is 2.18 (1.11) and for GooCard 1.43 (1.06). Thus, we can conclude that under optimal conditions, GooCard outperforms GooCost. For heuristics, both perform quite well. Using the estimated costs CF_{est} in BP_{trad} instead, these numbers increase to 3.15 (1.73) for GooCard with CE_{tru}. For BP_{smart}, the loss drops to 2.10 (1.32). It might not be intuitive why GooCard produces different loss factors for CF_{tru} and CF_{est}, although it only uses cardinalities and not costs for its join ordering decisions. In this particular case, the cost function is used by BP_{trad}. This also explains why there is no cost function shown for GooCard and BP_{smart}, as neither needs a notion of cost.

Turning to erroneous cardinality estimators for GooCard using BP_{smart}, we see that $CE_{IA\text{-}M}$ performs worse (15.53/1.94) than both CE_{base} (6.90/2.27) and CE_{sel} (6.71/2.32). No runtime-related cost function is needed here, similar to DPccp with BP_{smart} (last two lines of the DPccp-block). Comparing these, we see that going from DPccp to GooCard only slightly increases the maximum loss factor, while the average loss factor decreases for CE_{base} and remains about the same for CE_{sel}. CE_{base} has a slightly higher worst case than CE_{sel}.

Most DBMSs provide at least two different join ordering algorithms: one like DPccp for queries with moderate numbers of relations (say at most 15–20), and one heuristics for larger queries. The above numbers suggest that we can obviate DPccp and only implement GooCard in a first version of a newly developed DBMS without compromising performance too much, if we use CE_{base} or CE_{sel}. If we compare this scenario to the one where we implemented the cardinality estimator $CE_{IA\text{-}M}$ and some cost function CF_{est}, as well as the join ordering algorithm DPccp with the build-plan procedure BP_{trad}, we see that the combination of GooCard and BP_{smart} loses only a factor of about 2 in the worst case (going from 3.02 to 6.90/6.71 for CE_{base}/CE_{sel}) and a factor of $1.70 = 2.57/1.51$ for CE_{base} and $1.50 = 2.31/1.51$ for CE_{sel} on the average case.

VIII. RELATED WORK

Simplification of query optimizers is not a new idea. For example, Datta et al. propose the algorithm Simpli-Squared for join ordering without cardinality and cost estimation [2]. The basic idea is to first execute key/foreign key joins and then n:m-joins. Since no notion of cost/cardinality is available to Simpli-Squared, the ordering of relations for a star-query is random, depending on the order in which key/foreign key joins are enumerated (e.g., depending on the order of the relations in the `from`-clause).

Another example for query optimizer simplification is proposed by Hertzschuch et al. [3][25]. However, their approach highly intertwines the proposed cardinality estimation procedure with a newly proposed join ordering heuristics. Further, besides some cardinality estimates for filtered base relations, it also requires knowledge about the maximum multiplicity of the distinct values in the join attributes. Thus, it is much more complex than our cardinality estimators which are, in contrast, also independent of the join enumeration algorithm.

Notably, neither approach uses proper cost functions and leaves significant parts of the plan generation to PostgreSQL, relying on PostgreSQL's simple cost model whose errors remain unknown.

IX. Conclusion and Future Work

Since implementing and testing cost functions can be quite tedious, we showed that we can implement a competitive plan generator that does not rely on any cost function. Instead, the new build-plan procedure BP_{smart} makes all decisions based only on cardinality estimates. Further, we demonstrated that a very simple cardinality estimator CE_{base}, which does not even require cardinality estimation for single-table selection predicates, is quite competitive. Our evaluation shows that if the effort is undertaken to implement cardinality estimation for this case, e.g., based on sampling, then the average loss factor decreases when using CE_{sel}. Last but not least, we showed that implementing only GooCard and no other join ordering algorithm with optimality guarantees like DPccp results in only a limited loss of plan quality. Taking these findings together allows for an easy to implement query optimizer enabling a short time to market for a new DBMS.

For future work, we intend to perform an in-depth analysis on the granularity of individual queries and plans to extend our evaluation of plan quality.

Acknowledgments

We would like to thank Simone Kehrberg for proofreading the paper, and Nazanin Rashedi for helpful comments and discussions. We would also like to thank the anonymous reviewers for their suggestions to improve the paper.

References

[1] M. Raasveldt and H. Mühleisen, "DuckDB: An embeddable analytical database", in *Proc. of the ACM SIGMOD Conf. on Management of Data*, 2019, pp. 1981–1984.

[2] A. Datta, B. Tsan, Y. Izenov, and F. Rusu, "Simpli-squared: Optimizing without cardinality estimates", in *SiMoD '24: Proceedings of the 2nd Workshop on Simplicity in Management of Data*, 2024, pp. 1–10.

[3] A. Hertzschuch, C. Hartmann, D. Habich, and W. Lehner, "Simplicity done right for join ordering", in *Proc. Conference on Innovative Data Systems Research (CIDR)*, 2021.

[4] G. Moerkotte and T. Neumann, "Analysis of two existing and one new dynamic programming algorithm for the generation of optimal bushy trees without cross products", in *Proc. Int. Conf. on Very Large Data Bases (VLDB)*, 2006, pp. 930–941.

[5] G. Lohman, "Heuristic method for joining relational database tables", *IBM Technical Disclosure Bulletin*, vol. 30, no. 9, pp. 8–10, 1988.

[6] L. Fegaras, "Optimizing large OODB queries", in *Proc. Int. Conf. on Deductive and Object-Oriented Databases (DOOD)*, 1997, pp. 421–422.

[7] L. Fegaras, "A new heuristic for optimizing large queries", in *Int. Conf. on Database and Expert Systems Applications (DEXA)*, 1998, pp. 726–735.

[8] V. Leis *et al.*, "Query optimization through the looking glass, and what we found running the join order benchmark", *VLDB Journal*, vol. 27, pp. 643–668, 2018.

[9] G. Moerkotte, T. Neumann, and G. Steidl, "Preventing bad plans by bounding the impact of cardinality estimation errors", *Proc. of the VLDB Endowment (PVLDB)*, vol. 2, no. 1, pp. 982–993, 2009.

[10] G. Cormode, M. Garofalakis, P. Haas, and C. Jermaine, *Synopses for Massive Data: Samples, Histograms, Wavelets, Sketches* (Foundations and trends in databases). NOW Press, 2011, vol. 4:1–3, pp. 1–294.

[11] F. Olken and D. Rotem, "Simple random sampling from relational databases", in *Proc. Int. Conf. on Very Large Data Bases (VLDB)*, 1986, pp. 160–169.

[12] F. Olken, "Random sampling from databases", Ph.D. dissertation, U. California at Berkeley, 1993.

[13] P. G. Selinger, M. M. Astrahan, D. D. Chamberlin, R. A. Lorie, and T. G. Price, "Access path selection in a relational database management system", in *Proc. of the ACM SIGMOD Conf. on Management of Data*, 1979, pp. 23–34.

[14] D. Flachs, M. Müller, and G. Moerkotte, "The 3D Hash Join: Building On Non-Unique Join Attributes", in *Proc. Conference on Innovative Data Systems Research (CIDR)*, 2022.

[15] S. Chen, A. Ailamaki, P. B. Gibbons, and T. C. Mowry, "Improving hash join performance through prefetching", *ACM Transactions on Database Systems*, vol. 32, no. 3, 2007.

[16] O. Kocberber, B. Falsafi, and B. Grot, "Asynchronous memory access chaining", *Proc. of the VLDB Endowment (PVLDB)*, vol. 9, no. 4, pp. 252–263, 2015.

[17] Intel Corporation, *Intel 64 and IA-32 Architectures Software Developer Manual, Vol. 3B*, available at https://www.intel.com/content/www/us/en/developer/articles/technical/intel-sdm.html, Jun. 2024.

[18] S. Setzer, G. Steidl, T. Teuber, and G. Moerkotte, "Approximation related to quotient functionals", *Journal of Approximation Theory*, Special Issue: Bommerholz Proceedings, vol. 162, no. 3, pp. 545–558, 2010.

[19] G. Moerkotte, "Building query compilers", available at https://pi3.informatik.uni-mannheim.de/~moer/querycompiler.pdf, 2024.

[20] S. Cluet and G. Moerkotte, "On the complexity of generating optimal left-deep processing trees with cross products", in *Proc. Int. Conf. on Database Theory (ICDT)*, 1995, pp. 54–67.

[21] V. Leis *et al.*, "How good are query optimizers, really?", *Proc. of the VLDB Endowment (PVLDB)*, vol. 9, no. 3, 2015.

[22] G. Sun J. Li, "An end-to-end learning-based cost estimator", *Proc. VLDB Endow.*, vol. 13, no. 3, pp. 307–319, 2019. DOI: 10.14778/3368289.3368296.

[23] I. Trummer, "Exact cardinality query optimization with bounded execution cost", in *Proc. of the ACM SIGMOD Conf. on Management of Data*, 2019.

[24] R. Marcus *et al.*, "Neo: A learned query optimizer", *Proc. VLDB Endow.*, vol. 12, no. 11, pp. 1705–1718, 2019.

[25] R. Bergmann, A. Hertzschuch, C. Hartmann, D. Habich, and W. Lehner, "PostBOUND: PostgreSQL with upper bound SPJ query optimization", in *Proc. Fachtagung Datenbanksysteme für Business, Technologie und Web (BTW)*, 2023.

An Enhanced Semantic Framework for Time-Constrained Clinical Decision-Making in Emergency Settings

Sivan Albagli-Kim
Department of Mathematics and Computer Science
Seton Hall University
South Orange, NJ, USA
email: sivan.albaglikim@shu.edu

Dizza Beimel
Department of Computer and Information Sciences
Dror (Imri) Aloni Center for Health Informatics
Ruppin Academic Center
Emek Hefer 4025000, Israel
email: dizzab@ruppin.ac.il

Abstract— **Rapid and accurate decision-making is essential for identifying and treating life-threatening conditions in emergency medicine. This paper presents an enhancement to an existing Knowledge Graph-based clinical decision-making framework by integrating an emergency strategy layer to prioritize critical diagnoses. By categorizing diseases as life-threatening or non-life-threatening, our approach emphasizes the immediate exclusion of high-risk conditions. The enhancement is manifested on two primary levels: (a) we augmented the KG by incorporating conditional edges that are dynamically activated based on patient-specific indicators, such as age, gender, and pre-existing conditions. These conditional edges allow the framework to adapt to individual patient profiles, supporting a more precise and personalized diagnostic process; and (b) we refined the framework's algorithms to prioritize excluding life-threatening diseases. Future work will evaluate the framework with real-world clinical data and expand the KG's logic to include continuous data, further enhancing inference accuracy. Our contribution provides a foundation for expanding clinical decision-making frameworks to address urgent clinical needs, potentially improving patient outcomes in critical medical scenarios.**

Keywords- knowledge graph; semantic reasoning; decision support systems; semantic technology.

I. INTRODUCTION

Healthcare 4.0 addresses key challenges related to the expansion, virtualization, and innovation of modern healthcare practices, such as home-based care, precision medicine, and personalized or remote drug therapies [1]. It represents the shift towards leveraging advanced technologies to overcome barriers in healthcare delivery. In particular, we focus on utilizing semantic technologies powered by large datasets and complex algorithms.

Advancements in healthcare technology have increasingly leveraged Knowledge Graphs (KG) - a graph data model that has gained popularity for representing complex knowledge structures [2] - constructed from electronic medical records to enhance clinical decision support systems. Rotmensch et al. demonstrated the potential of such an approach by learning a health KG from electronic medical records, which improved the structuring of complex patient data and facilitated more accurate inferences in diagnostic processes [3].

This aligns with our framework's utilization of a KG to support medical experts in making timely and informed decisions, especially in environments where time constraints impact diagnostic accuracy. In our ongoing research [4][5] in the medical domain, we investigate clinical decision-making processes that facilitate interactions between healthcare experts and patients. The goal is to assist medical experts in helping patients resolve health issues. These interactions typically consist of multiple iterations, where the expert asks questions, and the patient responds. With each iteration, the medical expert moves closer to making a decision concerning the patient's condition, which usually culminates in a medical diagnosis. However, time constraints often limit these interactions, which can impact the accuracy of diagnoses.

To address the goal mentioned above, we developed a framework based on semantic technologies that support the decision-making process. Each iteration suggests a question relevant to the patient's symptoms. In the final iteration, the framework produces a ranked list of hypotheses consisting of disease-symptom pairs, ordered by the likelihood that the disease is the correct diagnosis. This framework is built on a KG, which effectively models interconnected data [6], with nodes representing symptoms and diseases and edges connecting symptoms to diseases when relevant. We have developed a set of interactive algorithms that utilize both the KG and the patient's initial input to suggest relevant questions during the interaction.

The basic KG was enriched with semantic knowledge, extracted from symptom ontology, expanded the knowledge base, and added hierarchic layers. The framework was fully implemented in Python and evaluated via a set of tests [5].

While the existing framework provides a solid foundation for supporting decision-making among medical experts, its scalability allows for easy extension through KG's capacity to grow in volume and knowledge layers; for instance, it can be enhanced by adding information to edges that can yield deeper insights and more accurate hypotheses. Based on this, our current research introduces an *emergency strategy knowledge layer* to enhance decision-making processes further. This layer simulates an emergency room environment, focusing on promptly identifying emergency conditions and appropriate treatment despite time-constrained communication between patients and medical experts. Interviews with two physicians revealed key insights: (1) physicians observe additional symptoms through physical examinations and abnormal vital signs beyond those reported by patients; (2) personal patient information (e.g., age, gender, pre-existing conditions, medications) is crucial in diagnostics; (3) diseases are classified into life-threatening and non-life-threatening categories based on symptoms and patient data; and (4) the proposed strategy prioritizes the immediate exclusion of life-threatening conditions.

Following the above, the basic architecture of our framework has been enhanced to provide medical experts with a list of hypotheses related to life-threatening diseases. This enhancement includes augmenting the KG with conditional edges based on patient-specific indicators (such as age, gender, and pre-existing conditions) and refining the framework algorithms to prioritize excluding life-threatening diseases. The list of hypotheses is generated through an inference process that searches for symptoms to either rule out or confirm life-threatening conditions. By simulating an emergency room environment, this enhancement enables the framework to prioritize the rapid identification of critical conditions in time-sensitive settings.

The paper is organized as follows: Section 2 discusses knowledge representation and reviews studies that utilize KGs for healthcare applications. Section 3 details our framework, and Section 4 describes the framework enhancements developed to support the emergency strategy. Finally, Section 5 concludes with a summary of contributions and suggestions for future work.

II. Background

In this section, we discuss how knowledge can be represented and provide an overview of researches that use KG in healthcare-related applications.

A. Knowledge Representation

Knowledge Representation (KR) serves several essential roles, such as enabling entities to predict the outcomes of actions, establishing frameworks for perceiving the world, providing foundations for intelligent reasoning, facilitating efficient computation, and acting as a medium for human expression [7]. Key methods of KR include KGs, ontologies, and semantic technologies.

KGs, also known as semantic graphs, represent information by encoding relationships between entities into graph structures. They offer semantically structured data that supports innovative solutions in tasks like question answering, recommendation systems [8], and information retrieval [9]. KGs hold significant promises for developing more intelligent machines.

Ontologies are explicit, machine-interpretable specifications of conceptualizations, defining entities within a domain, their attributes, and their interrelationships [10]. They establish a common vocabulary for humans and machines to share information, facilitating shared understanding, reuse of domain knowledge, and systematic analysis [11].

Semantic technologies aim to derive meaning from information by managing knowledge and integrating diverse data streams for inference. By representing both data and domain knowledge using graph models—since ontologies are often graph-based—graph algorithms can be employed to infer new insights.

B. Literature Review

Recent advancements in clinical decision support systems have increasingly leveraged KGs to enhance diagnostic accuracy and personalized care. For example, the construction and evaluation of causal KGs for diabetic nephropathy have demonstrated improved support in clinical decisions by modeling complex causal relationships within patient data [12]. Similarly, integrating KGs with large language models has been explored to enhance emergency decision-making, providing real-time support in critical care scenarios [13]. Furthermore, incorporating proteomics data into clinical decision-making through clinical KGs has shown promise in personalized medicine, allowing for more precise diagnostics and tailored therapies [14]. Additionally, enriching KGs from clinical narratives using natural language processing (NLP), Named Entity Recognition (NER), and biomedical ontologies has advanced healthcare applications by improving the extraction and structuring of valuable clinical information [15]. These studies underscore the significant potential of KGs in augmenting clinical decision support systems, particularly when combined with semantic technologies and patient-specific data. This aligns with our approach of integrating an emergency strategy layer into a KG-based framework to prioritize life-threatening conditions.

Recent studies have applied machine learning to clinical decision support, relying on large datasets for diagnosis prediction. While effective, these methods often require extensive labeled data and lack interpretability. Our approach, based on semantic reasoning within a KG, enables transparent and adaptive decision-making, allowing experts to incorporate new insights in real-time without retraining, enhancing explainability in time-sensitive medical settings.

III. The Framework

This section summarizes the framework we developed in our previous study [4][5], detailing its key algorithms and how they interact.

Recall that our objective is to support collaborative decision-making through an efficient exchange between a domain expert and an end-user, where both parties share questions and answers. In the medical context, the domain expert is a medical expert, and the end-user is a patient. The questions and answers revolve around symptoms and potential diagnoses. The framework facilitates this interaction by suggesting relevant questions for the medical expert to ask the patient (e.g., "Does the patient exhibit a particular symptom?"), with the decision-making process advancing based on the patient's responses. The framework output is a ranked list of hypotheses, where each hypothesis links a specific disease to a related symptom. As a result, the key terms in this framework are *symptoms*, *diseases*, and *hypotheses*.

The framework utilizes a KG, a widely adopted approach for representing knowledge [5]. KGs have become increasingly popular due to their ability to represent interconnected data [16][17] naturally. In this context, the KG comprises `nodes` representing `symptoms` and `diseases`, with edges connecting a symptom to a disease when the symptom indicates that condition, named `symptomOf` within the KG. Building on the KG, we formulated an inference process comprised of a set of developed interactive algorithms that leverage both the KG and the patient's initial input to generate relevant questions for the medical expert.

The framework comprises two main stages: (1) an initial pre-processing phase upon framework initialization and 2) a subsequent processing phase triggered with each new patient interaction.

A. Pre-Processing Phase

In the pre-processing phase, a KG is constructed from raw data from Kaggle [18] using Neo4j Graph Database, Version 5 [19]. The dataset consists of patient records, each corresponding to a single patient. These records include each patient's diagnosed disease and the symptoms they reported. In total, the dataset covers 41 distinct diseases and 130 unique symptoms. Some symptoms appear only once, indicating they are linked to a single disease, while others are associated with multiple diseases.

The KG was enriched by semantic knowledge extracted from an ontology of symptoms (SYMP) [20] and their interrelationships. Key elements from this ontology, particularly its hierarchical structures, were incorporated into the KG as follows: the symptoms were defined in the KG as `ontology symptoms`, and the hierarchical relationships were defined as `ISA` edges. The enriched KG, with its expanded symptom representations and hierarchical organization, offers several advantages for the inference process. These enhancements include a wider range of recommended questions for the medical expert during each interaction with the patient and symptoms that can be represented by the patient (referred to as *evidence symptoms* or ES) [21]. Figure 1, a Neo4j screenshot, demonstrates a subgraph of the enriched KG, particularly the creation of the `cough` symptom node, which is linked by a `symptomOf` edge to the `GERD` disease node. Additionally, it shows the node for its descendant (e.g., `dry cough`), connected to the parent node via an `ISA` edge. Note that the `dry cough` node has its descendant, the `dry hacking cough` node.

Finally, we applied the Louvain hierarchical clustering algorithm [22] to the KG to identify clusters of diseases—called *communities*—that share similar symptoms. We named this step as Algorithm 1 [3].

B. Processing Phase

The processing phase begins whenever a new interaction between a medical expert and a patient starts, with the patient presenting evidence of symptoms (ES). During this interaction (named Algorithm 2 [3]), the framework executes inference algorithms that utilize the identified communities to determine which diseases are compatible with the patient's symptoms. Specifically, Algorithm 2 identifies the most probable diseases that align with the evidence symptoms. Next, Algorithm 3 [3] iteratively, as needed, suggests to the medical expert questions (i.e., symptoms) that point toward the community most likely to include the patient's disease. Finally, the processing phase concludes with Algorithm 4 [3], which infers and outputs a ranked list of hypotheses—ordered pairs of a disease and an indicative symptom—that the patient might have.

The entire framework was implemented in Python, and we conducted a series of tests to evaluate its output and effectiveness [4].

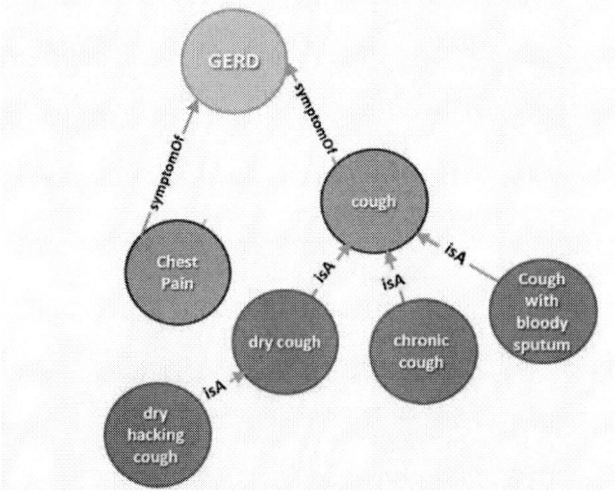

Figure 1. An example of integrating a hierarchical tree of symptoms into the KG. Disease nodes are represented in gray, symptom nodes in yellow, and ontology nodes in red.

IV. Extending the Framework: Integrating an Emergency Strategy

Within this section, we describe the framework enhancement we developed to support an emergency strategy along with its formal representation and its set of algorithms. In addition, we provide two simple examples that demonstrate how the enhanced framework can be utilized in emergency mode.

A. Motivation

The existing framework supports decision-making processes and is easily extendable through a scalable KG) that can incorporate additional insights. Our current research adds an emergency strategy knowledge layer to the KG, simulating an emergency room setting to prioritize identifying and treating life-threatening conditions under time constraints.

To develop this emergency strategy, we interviewed two physicians and gathered the following key insights:

1. In addition to the symptoms reported by the patient, there are other symptoms observed by the physician, which result from physical examination and abnormal vital signs (e.g., blood pressure outside the normal range).
2. Personal information about the patient (in particular, age, gender, pre-existing conditions, and medications) plays a crucial role in the diagnostic process.
3. given the symptoms and the above data, the possible diseases are classified into two categories: life-threatening and non-life-threatening.
4. The proposed strategy: first, rule out immediate life-threatening conditions.

In the following subsections, we describe the strategy and how we formulated it into a representation and a set of algorithms integrated into our framework.

B. Emergency Strategy Overview

To incorporate the aforementioned insights, we undertook two main actions: (a) enhancing the KG and (b) enhancing the processing phase to support the emergency strategy.

1) KG Enhancement

KG enhancement involves two steps, both conducted during the pre-processing phase:

A. Risk Attribute for Diseases: For all diseases in the graph, we add a Boolean attribute called *risk?*, which indicates whether a disease needs to be ruled out promptly or not.

B. Incorporating three indicators: age, gender, and pre-existing conditions into the KG. To incorporate the influence of these indicators on the presence of a life-threatening disease, we define a new type of `SymptomOf` edges: *conditional edges*. These edges are associated with an attribute formulated as a logical rule. The rule can involve one to three indicators connected with AND/OR operators.
For instance, to signal a life-threatening condition given a symptom *s1* indicating disease *d1*, if the patient is over 60 years old and male, the rule would be formulated as (*age > 60 AND gender = M*), and it assigns the conditional `SymptomOf` edge from *s1* to *d1*.

The second step (B) involves categorizing the `SymptomOf` edges in the KG into unconditional and conditional edges. Unconditional edges represent permanent relationships between symptoms and diseases that are universally applicable. In contrast, conditional edges are associated with logical rules involving patient-specific indicators. These conditional edges are incorporated into the patient's graph at runtime (processing phase) only when their associated logical rules are evaluated to be *true*. This mechanism allows the graph to adjust to individual patient profiles dynamically, enabling more precise and personalized inference during the diagnostic process. Figure 2 presents an example of a KG that was enhanced according to the described steps: It includes two diseases (*d2*, *d5*) that are characterized as high-risk, and conditional edges (e.g., the edge from *s5* to *d3*, marked with "*age<2*").

2) Processing Phase Enhancement

Enhancing the processing phase builds upon the original process by introducing new algorithms that support the emergency strategy. The evidence input process has expanded to include, beyond the symptoms reported by the patient, vital signs (such as blood pressure), and additional symptoms discovered by the medical expert during the patient's examination (e.g., a rigid abdomen). Despite the broader range of evidence entering the framework, all inputs are still characterized as *evidence symptoms*. Additionally, the patient's data, specifically the three noted indicators (age, pre-existing conditions, gender), are inputted. At this point, a new algorithm is introduced, which performs logical inference on the dependent edges. If the logical rule evaluates to true for each such edge, the associated edge is added to the patient's graph.

With the patient's graph now prepared for further analysis, identifying possible diseases and inferring potential communities proceeds with a slight modification to Algorithm 2: it now sorts the possible diseases as follows: a primary sorting of diseases with the attributed *risk? = true*, followed by a secondary sorting of all other diseases. Subsequently, the communities are ranked based on their disease scores. The disease score is decided, as before, by the degree of supporting evidence, that is, the number of evidence symptoms pointing to the disease, including the conditional edges becoming *true* (e.g., if three evidence symptoms indicate a disease, its score is 3).

Algorithms 3 and 4 are executed as described in [4][5]. Thus, for each community, we search for a symptom that can either rule out or confirm a life-threatening disease and the inference process concludes with a ranked list of hypotheses that the patient might have. Naturally, if the inference process identifies any life-threatening diseases, they will be prioritized first.

C. Formalizing the Emergency Strategy

In this section, we provide a formal description of how the strategy aligns with the KG, which includes the refined KG process and is supported by the algorithms.

1) KG and Pre-processing Formalizing

Refining the KG includes two main steps, as explained earlier. Both steps are implemented in the framework's pre-processing phase, as they do not involve the patient and remain consistent across patients.

A. Identify the diseases with high risk and add a Boolean attribute that recognizes them in the graph:
   a. Let D be the set of nodes representing the diseases in the KG. For every disease $d \in D$, add a Boolean attribute named risk? with the default value *false*.
      Let $D_{risk} \subseteq D$ Be the set of diseases with high risk. For each disease $d \in D_{risk}$, set risk? to *true*.

B. Incorporating the indicators age, gender, and pre-existing conditions into the KG: this step translates a set of rules R into conditional edges E_C in the processing step. Each rule $r \in R$ represented by a tuple $\langle s, d, f(i_{age}, i_g, i_{pre})\rangle$, where s is a symptom, d is a disease, and f is a boolean function that receives three personal indicators (i_{age}, i_g, i_{pre}) representing age, gender, and pre-existing conditions, respectively. The function f returns *true* if s indicates d according to the patient indicators. The set of conditional edges E_C are defined as follows: $E_C = \{(s,d) | f(i_{age}, i_g, i_{pre}) = true\}$. These edges will be evaluated during the processing step when a patient arrives.

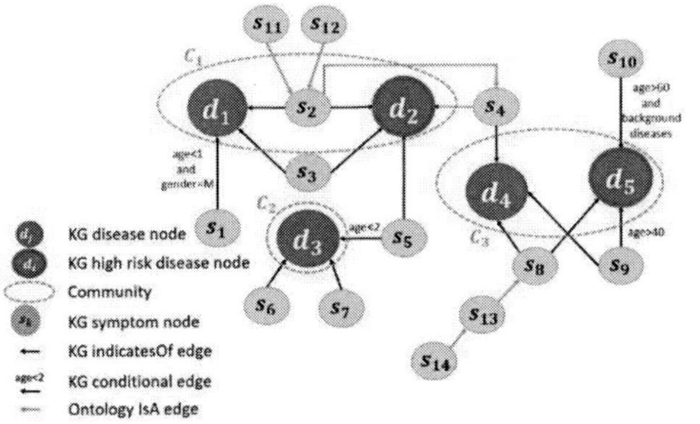

Figure 2. The Enhanced KG

2) Framework-specific Terminology

Table 1 (An extensive view is in Appendix 1) presents the terminology that we use to describe the algorithms. Additional terms supporting the emergency strategy are bold.

3) The Refined Framework Algorithms

We describe the refined algorithms developed in our framework to support the emergency strategy.

Algorithm 1 builds the personalized subgraph from KG by adding the patient's personal information (the indicators). Algorithm 2 incorporates the patient's symptoms into the personalized sub-graph and uses inference to generate a ranked list of potential diseases. This list then serves as the input for Algorithms 3 and 4 [4][5], which return a set of hypotheses prioritized by their urgency. Each hypothesis is a pair consisting of a disease and a symptom indicating it.

Algorithm 1: personalized sub-graph
Input: $KG = (DUS, E)$, PI, ES
Output: personalized sub-graph PKG
Algorithm:
0. Let PKG be KG
1. For every $s \in ES$:
   a. For every $r \in R$ that contain s, that is $r = \langle s, d, f\rangle$:
   b. If f(PI)=true, add the edge (s, d) to PKG.
2. Return PKG

Figure 3. Algorithm 1: personalized sub-graph

Algorithm 2: identify possible diseases
Input: PKG, ES, C
Output: possible diseases, sorted according to their risk
Algorithm:
1. Let $PD \leftarrow \{\}$
2. Let $C' \subseteq C$ be the set of communities having positive LinD.
3. Sort C' in a non-decreasing order according to their Risk (primary), and then according to their LinD (secondary).
   3.1. Let c be the community in the order:
      3.1.1 Go over the diseases in c.
         First go over the diseases d with risk?==true. Sort them according to their R^d(d) (in a decreasing order) and add them in that order into PD.
         Then add the rest of the diseases in c, sorted (in a decreasing order) according to their R^d(d).
4. return PD

Figure 4. Algorithm 2: identify possible diseases (sorted according to risk)

D. Simplified Example

We illustrate two distinct scenarios involving different patients who exhibit the same symptoms: s1, s5, s9, and s10. However, despite sharing identical symptoms, the patients in each case have unique personal characteristics and health profiles.

The first scenario involves a 75-year-old man with no prior health conditions. The resulting graph (PKG1), after his personal indicators were entered and processed, is presented in Figure 5.

The second scenario involves a 9-month-old baby without any prior health conditions. The resulting graph (PKG2) after inputting and processing his personal indicators is shown in Figure 6.

It is important to note that these two scenarios produce different graphs, meaning the algorithms process different inputs and generate distinct hypotheses. In the first scenario, only communities *C1* and *C3* are examined, and since $Risk(C3) \geq Risk(C1)$, the first disease to be ruled out is *d5*. In the second scenario, all communities are considered, and since: $Risk(C1) \geq Risk(C2) = Risk(C3)$, the first disease to rule out is *d2*.

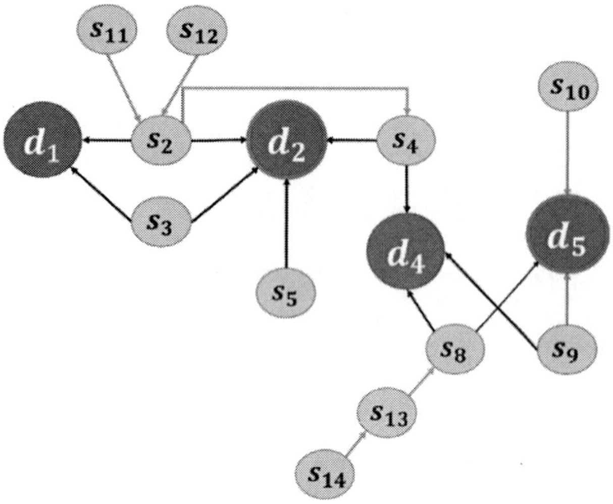

Figure 5. PKG1 – the graph for the 75-year-old man

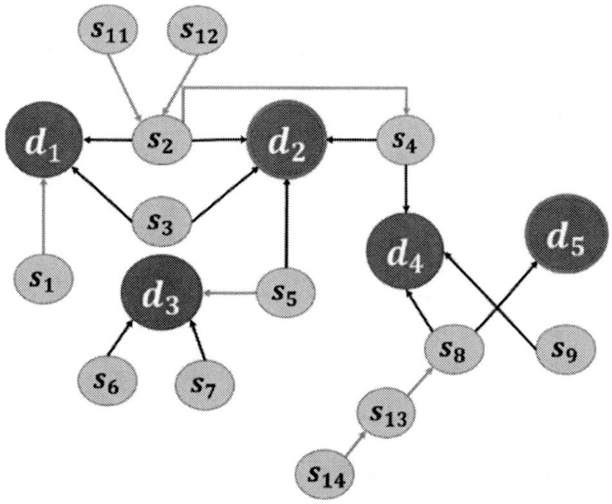

Figure 6. PKG2 – the graph for the 9-month-old baby

V. Conclusion and Discussion

In emergency medicine, the rapid identification and treatment of critical conditions are essential for optimizing patient outcomes. Horng et al. demonstrated the effectiveness of machine learning in developing an automated trigger for sepsis clinical decision support at emergency department triage, showcasing how advanced technologies can enhance patient care in high-stakes settings [23]. Our enhancement similarly integrates an emergency strategy layer into our KG-based decision-making framework to prioritize life-threatening conditions, thus improving decision-making efficiency and patient outcomes in urgent scenarios.

A. Summary

This study categorizes diseases based on symptoms and patient data into two main types: life-threatening and non-life-threatening. The proposed strategy focuses on the rapid exclusion of life-threatening diseases, which is crucial for optimizing patient outcomes in emergency care.

To improve the inference process for identifying life-threatening conditions, we augmented the KG by incorporating *conditional edges*. These edges, which rely on patient-specific indicators such as age, gender, and pre-existing conditions, are dynamically added to the patient's graph at runtime when specific conditions are met. This adaptive approach allows the decision support framework to tailor diagnostics to individual patient profiles, facilitating more precise and personalized recommendations.

B. Contribution

Our work advances clinical decision-making processes by formulating and integrating an emergency strategy prioritizing life-threatening conditions. We developed an enhanced KG with conditional edges informed by patient-specific data, allowing for real-time personalization. We also refined existing algorithms to incorporate this emergency strategy, enabling a diagnostic process that is more accurate and responsive to critical clinical needs. These contributions establish a more adaptable decision-making framework for emergency contexts, providing a robust foundation for further developments in emergency medical diagnostics.

C. Next Phase and Future Work

The next phase of this research will involve validating the emergency strategy using real-world clinical data to assess its effectiveness in supporting healthcare professionals in practice. Furthermore, we plan to refine the logic for conditional edges by incorporating continuous data, which will improve inference granularity and diagnostic accuracy within the KG. Expanding this work, we aim to integrate machine learning models that dynamically update the KG based on incoming data, thereby increasing the system's adaptability to evolving clinical practices and patient populations.

REFERENCES

[1] G. Aceto, V. Persico, and A. Pescapé, "Industry 4.0 and health: Internet of things, big data, and cloud computing for healthcare 4.0.," *Journal of Industrial Information Integration*, vol. 18, 100129, 2020.

[2] R. J Webber and E. Eifrem, *Graph Databases: New Opportunities for Connected Data*, O'Reilly Media, Inc.: Middlesex County, MA, USA, 2015.

[3] M. Rotmensch, "Learning a health knowledge graph from electronic medical records," *Scientific reports* vol. 7, no. 1, pp. 5994, 2017.

[4] S. Albagli-Kim and D. Beimel, "Knowledge graph-based framework for decision-making process with limited interaction," *Mathematics*, vol. 10, no. 21, pp. 3981, 2022.

[5] D. Beimel and S. Albagli-Kim, "Enhancing Medical Decision Making: A Semantic Technology-Based Framework for Efficient Diagnosis Inference," *Mathematics*, vol. 12. No. 4, pp. 502, 2024.

[6] E. Rajabi and S. Kafaie, "Knowledge graphs and explainable AI in healthcare," *Information*, vol. 13, no. 10, pp.459, 2022.

[7] R. Davis, H. Shrobe, and P. Szolovits, "What is a Knowledge Representation?," *AI Magazine*, vol. 14, no. 1, pp. 17-33, 1993

[8] Q. Guo, et al., "A survey on knowledge graph-based recommender systems," *IEEE Transactions on Knowledge and Data Engineering*, vol. 34, no. 8, pp. 3549–3568, 2020.

[9] L. Dietz, A. Kotov, and E. Meij, "Utilizing knowledge graphs for text-centric information retrieval." *The 41st International ACM SIGIR Conference on Research & development in Information Retrieval*, pp. 1387–1390, 2018.

[10] T. R. Gruber, "Toward principles for the design of ontologies used for knowledge sharing?," *International journal of human-computer studies*, vol. 43, no. 5-6, pp. 907-928, 1995.

[11] N. F. Natalya and D. L. McGuinness, "Ontology development 101: A guide to creating your first ontology," 2001.

[12] K. Lyu, et al., "Causal knowledge graph construction and evaluation for clinical decision support of diabetic nephropathy," *Journal of Biomedical Informatics*, 139: 104298, 2023.

[13] M. Chen, et al., "Enhancing emergency decision-making with knowledge graphs and large language models," *International Journal of Disaster Risk Reduction*, 113: 104804, 2024.

[14] A. Santos, et al., "Clinical knowledge graph integrates proteomics data into clinical decision-making," bioRxiv (2020), 2020.

[15] A. Thukral, S. Dhiman, R. Meher, and P. Bedi, "Knowledge graph enrichment from clinical narratives using NLP, NER, and biomedical ontologies for healthcare applications," *International Journal of Information Technology*, vol. 15, no. 1, pp. 53-65, 2023.

[16] M. Besta, et al., "Graph of thoughts: Solving elaborate problems with large language models," *Proceedings of the AAAI Conference on Artificial Intelligence*, Vol. 38, No. 16, pp. 17682-17690, 2024.

[17] Q. Wang et al., "Knowledge graph embedding: A survey of approaches and applications," *IEEE Transactions on Knowledge and data engineering*, vol.29, no. 12, pp. 2724-2743, 2017.

[18] Kaggle. Available online: https://www.kaggle.com [retrieved June, 2024].

[19] Neo4j. Available online: https://neo4j.com/ [retrieved: January 2025].

[20] Symptom Ontology (SYMP), Ontology Lookup Service (OLS). Available online: https://www.ebi.ac.uk/ols/ontologies/symp, [retrieved: January, 2025].

[21] S. Bonner, et al., "Understanding the performance of knowledge graph embeddings in drug discovery," *Artificial Intelligence in the Life Sciences*, vol.2: 100036, 2022.

[22] H. Lu, M. Halappanavar, and A. Kalyanaraman, "Parallel heuristics for scalable community detection," *Parallel Computing*, vol. 47, pp. 19-37, 2015.

[23] S. Horng, et al., "Creating an automated trigger for sepsis clinical decision support at emergency department triage using machine learning," *PloS one*, vol.12, no. 4: e0174708., 2017.

APPENDIX 1

TABLE 1: THE EXTENDED ALGORITHMS TERMINOLOGY

Term	Definition
D	*The set of disease nodes*
$\boldsymbol{D_{risk}}$	***Let*** $\boldsymbol{D_{risk} \subseteq D}$ ***be the set of high-risk diseases.***
S	*The set of symptom nodes*
ES	*The set of evidence symptoms (i.e., the symptoms indicated by the patient)*
PI	***The patient's personalized indicators***
C	*The set of communities*
$\lvert c \rvert$	*The size of a single community* $c \in C$. *Defined by the number of diseases that belong to c*
Risk(c)	***Defined by diseases number of diseases in*** $\boldsymbol{D_{risk}}$ ***+ the number of evidence symptoms indicates a dieases in*** $\boldsymbol{D_{risk}}$
LinD(c)	*The Local-in-Degree of a given* $c \in C$. *Defined by the number of edges that point to diseases of c, by ES, hence, it is the sum of* $R^c(s,c)$, *for each* $s \in ES$ *and the given c*
PD's communities	*The set of communities* $c \in C$ *with a positive LinD(c), hence, a community in which at least one edge from* $s \in ES$ *points to c*
$R^d(d)$	*The Disease's Symptoms Rank.* *Defined by the number of symptoms the patient has that indicate D*

A Low-Code Approach for Creating Dynamic Map-Based Web Applications Using W3C Web Components

Andreas Schmidt*‡ and Tobias Münch†
* University of Applied Sciences
Karlsruhe, Germany
Email: andreas.schmidt@h-ka.de
‡ Karlsruhe Institute of Technology
Karlsruhe, Germany
Email: andreas.schmidt@kit.edu
† Münch Ges. für IT Solutions mbH, Germany
Email: to.muench@muench-its.de

***Abstract*—We present a set of web components that enable the declarative development of web-based, map-centered applications in a simple way. To do this, we used the World Wide Web Consortium (W3C) standard *Web Components*, which enables the development of new HTML elements. The developed components encapsulate the functionality of the *leaflet library*, a widely used Javascript library for the realization of map-based applications. The declarative approach makes it possible for non-programmers to develop applications within a short time. Some of the developed components have interfaces for accessing server-side information, such as *GeoJSON*-data sources and time series databases. This makes it possible to develop information-rich, "live" applications with our components.**

Keywords-dynamic interactive map applications; low-code; web-components; geojson; time series databases.

I. Introduction

For geospatial data, maps are the ideal form of presentation. The Open Street Map initiative [1] has created a global, freely available dataset that is suitable for displaying electronic maps of any scale. *Leaflet* [2] is a popular JavaScript library for developing web-based map applications that run in the browser.

A. *Leaflet*

The main visual components of the *leaflet library* are basemaps, overlay-layers, and geometric elements, such as markers, lines, polylines, circles, polygons (closed polylines), and rectangles.

The map displayed is composed of a basemap and optional overlay-layers. There are many freely available basemaps for *leaflet*, many of which are based on the open street map dataset [1], but there are also basemaps from other providers, such as Google, ESRI, etc. Overlays consist of images (i.e., png, svg) that are placed on top of the baselayer and enrich it with additional information, such as nautical marks [3] or hiking routes. Overlays can be shown or hidden on the map as needed.

Markers represent specific points on a map and can be displayed either with a standard visualization or with your own icon. The graphical elements can be enriched with *tooltips* and *popup menus*. In addition, leaflet has a wide range of events to which you can attach your own functions.

Graphic components, such as markers, lines and polygons can be grouped in so-called *layer-groups*. The idea behind this is that these can then be easily shown or hidden with a single instruction, analogous to the overlays.

The GeoJSON class allows the display of graphical features that are described in the form of GeoJSON [4] data sets. These can be points, (multi) lines, (multi) polygons, or geometric collections (multiple instances of the previous types).

B. *Web-Components*

The leaflet library is one of the most widely used JavaScript libraries for developing map-based applications. Its advantage lies in its technical maturity and good documentation. Nevertheless, the library requires in-depth knowledge of the underlying classes and programming experience with JavaScript to develop even the simplest applications. In contrast, this work takes a low-code approach in which the components of the card-based application are defined declaratively. The W3C composite standard *web-components* [5] is used for this purpose. A web component is a JavaScript class that must implement a series of methods in order to be integrated into the Document Object Model (DOM) tree within a website. The class is then mapped to an HTML tag-name that can then be used within the website.

The main goal of our work is to provide the leaflet library with a new declarative interface based on the W3C standard Web Components, so that non-programmers are also able to create map-based applications. But our components also make things easier for the experienced developer, since a large part of the code that would typically need to be implemented can be covered by our components, and only specific parts need to be implemented by hand. This is possible because the leaflet objects encapsulated by the *Web Components* can be accessed at any time.

The remainder of the paper is organized as follows: In Section II, some related work will be presented. Section III presents the overall architecture before Section IV demonstrates an example application that shows the implemented components in action. Section V concludes the paper with a summary and an outlook on further work in this field.

II. Related work

An approach using the Polymer [6] library for building a web-component interface around the leaflet library was established in 2015 [7]. However, the development seemed stopped in 2016. We follow the approach chosen there in the structural design, but go further in that our components access data provider on the server side. This can be used in particular to implement "live scenarios", as well as to load and display GeoJSON objects from multiple data-sources. Furthermore, the development of leaflet has continued over the last 9 years and our components also use additional packages, such as hierarchical clustering of markers at larger scales to avoid cluttering the map view [8]. The DB-Web Components [9] that we developed last year follow an analogous low-code approach in which relational database content is embedded declaratively in HTML pages.

III. Architecture

The architecture of our application is shown in Figure 1. On the client side, our components run within a browser. There they are responsible for displaying maps, overlays, and geographic features, such as markers, lines, etc. A number of components can optionally obtain their information from data sources on the server side. For example, the marker component can cyclically read its position from a time series database, or our GeoJSON component can access one or more `FeatureSets` and thus visualize many objects of different types with a single instruction.

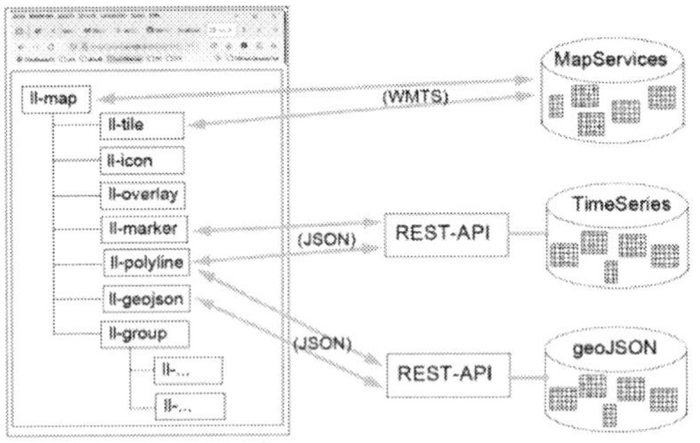

Figure 1. Architecture.

Next, we will show by way of example how the components can be used to implement a sample application.

IV. Example

Figure 2 shows the implemented application. On it, two paddlers can be seen paddling towards each other to enjoy a beer together at the picnic spot (represented by a beer icon). The associated application code can be seen in Figure V. The basemap and an additional alternative ESRI satellite image are defined between line 2 and line 12 in the context of the `ll-map`, `ll-tile`, and `ll-overlay` component. The `ll-overlay` element starting at line 9 defines that an additional overlay, labeled "Sea layer", with nautical symbols is displayed on top of the basemap.

The `ll-geojson` element on line 13 loads the data about the portages (location where the boat has to be transported over land) from a server-side `geojson` data source specified by the `url` parameter. In addition, an icon symbol (a paddler carrying his boat overhead) is specified using the `icon-url` and `icon-size` parameters.

In line 19, the picnic spot is defined at a fixed point specified by longitude and latitude. It also has its own icon and a tooltip text that is displayed when the icon is clicked.

The two paddlers to be displayed are defined in lines 25 and 31. Instead of hard-coding the position of the two paddlers as at the picnic area, you can get their current position cyclically from a time series database, which is specified by the parameter `url`. The two further parameters `lat-path` and `lng-path` specify where the information about longitude and latitude is located in the *json* response.

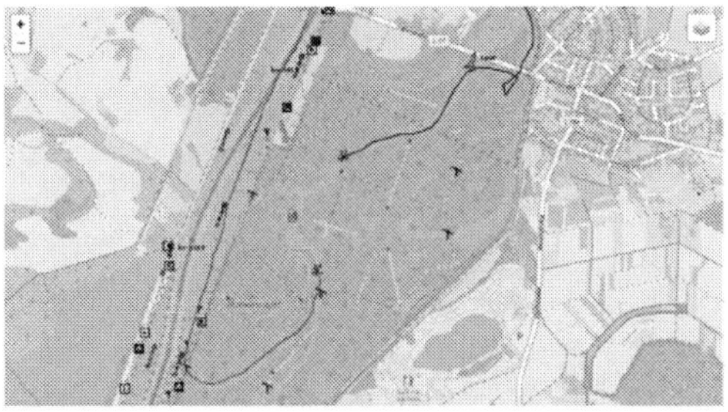

Figure 2. Screenshot of Example Application

Finally, the element `ll-group` is to be explained. As the name suggests, it functions as a grouping element and groups the child elements, which can then be switched on and off using the *layer control* at the top right. This can be seen in Figure 3, where the opened layer control is visible at the top right.

Figure 3. Open *layer control* (top right), which handles the selection of the basemap ("Esri Satelitte"), as well as the overlays ("Sea layer") and groups ("Portages [Img]", "Paddle tour ...") to be shown.

In the figure, the alternative Esri satellite image was selected as the basemap and the *sea layer* with the nautical signs is turned off. The `ll-geojson` data source (portages) also appears in the layer control under the specified label.

V. Conclusion and Outlook

We have implemented a first prototype based on the Lit Framework [10]. In addition to the components presented in the example, there is also an `ll-polyline` component for displaying lines and an `ll-icon` component for the

```
<body>
  <ll-map id="mymap"
          zoom="15">
    <ll-tile url='http://server.arcgisonline.com/ArcGIS/rest/services/World_Imagery/\
                  MapServer/tile/{z}/{y}/{x}'
                  label="Esri␣Satellite">
             &copy; <a href="http://www.esri.com/">Esri</a>
    </ll-tile>
    <ll-overlay url="http://t1.openseamap.org/seamark/{z}/{x}/{y}.png"
                label="Sea␣layer">
             &copy; <a href="http://t1.openseamap.org/copyright">OpenSeaMap</a>
    </ll-overlay>
    <ll-geoJSON label="Portages␣<img␣src='icons/portage.png'␣width='15'>"
                icon-url="./icons/portage.png"
                url="http://localhost/llwc/readGeoJSON.php?file=data/portages.json">
                path="result">
    </ll-geoJSON>
    <ll-group label="Paddle␣tour␣February␣22,␣2024">
      <ll-marker lat="48.8776"
                 lng="8.1339"
                 icon-size="20"
                 icon="icons/beer.png"
                 tooltip="Picnic<br>Place">
      </ll-marker>
      <ll-marker icon="icons/kanu2.png" icon-size="20"
                 tooltip="Thomas"
                 url="http://localhost/llwc/readFromInflux.php/thomas?num=1"
                 lat-path="result[0].lat"
                 lng-path="result[0].lon">
      </ll-marker>
      <ll-marker icon="icons/kanu.png" icon-size="20"
                 tooltip="Andreas"
                 url="http://localhost/llwc/readFromInflux.php/andreas?num=1"
                 lat-path="result[0].lat"
                 lng-path="result[0].lon">
      </ll-marker>
    </ll-group>
  </ll-map>
</body>
```

Figure 4. Markup code for example application

definition of icon objects, which can then be referenced by other components.

Further work is planned in the area of simplified integration of JavaScript functions into the leaflet event mechanism, as well as the processing of feature information when representing *GeoJSON* objects.

References

[1] OpenStreetMap, "OpenStreetMap", Last accessed 17.1.2025, 2024, [Online]. Available: https://www.openstreetmap.org.

[2] V. Agafonkin, "Leaflet api reference", Last accessed Last accessed 17.1.2025, 2024, [Online]. Available: https://leafletjs.com/reference.html.

[3] OpenSeaMap, "The free nautical chart", Last accessed 17.1.2025, 2024, [Online]. Available: https://openseamap.org/index.php?id=openseamap&L=1.

[4] H. Butler, S. Gillies, and T. Schaub, "Rfc 7946 - the geojson format", Last accessed 17.1.2025, 2016, [Online]. Available: https://datatracker.ietf.org/doc/html/rfc7946.

[5] webcomponents.org, "WebComponents - Specifications", Last accessed 17.1.2025, 2024, [Online]. Available: https://www.webcomponents.org/specs.

[6] POLYMER, "Polymer library", Last accessed 17.1.2025, [Online]. Available: https://polymer-library.polymer-project.org/.

[7] Hendrik Brummermann et al., "Leaflet-map", Last accessed 17.1.2025, 2015, [Online]. Available: https://github.com/leaflet-extras/leaflet-map.

[8] Erik Nikulski, "Leaflet.markercluster", Last accessed 17.1.2025, 2024, [Online]. Available: https://github.com/Leaflet/Leaflet.markercluster.

[9] A. Schmidt and T. Münch, "Web Components for Database Developers", in *Proceediungs of the Sixteenth International Conference on Advances in Databases, Knowledge, and Data Applications*, (Athen, Griechenland, Mar. 10–14, 2024), 2024, pp. 20–22.

[10] LIT, "Simple. Fast. Web Components", Last accessed 17.1.2025, 2024, [Online]. Available: https://lit.dev/.

Decentralized Browser-based Cloud Storage: Leveraging IPFS for Enhanced Privacy

Georg Eilnberger
St. Pölten UAS
St. Pölten, Austria
e-mail: is211806@fhstp.ac.at

Timea Pahi
St. Pölten UAS
St. Pölten, Austria
e-mail: timea.pahi@fhstp.ac.at

Peter Kieseberg
St. Pölten UAS
St. Pölten, Austria
e-mail: lbkieseberg@fhstp.ac.at

***Abstract*—This work addresses the growing need for data management solutions that prioritize security, privacy, and user control, amidst the limitations of traditional centralized storage systems with a particular focus on the InterPlanetary File System (IPFS). The core objective is to explore the efficiency, challenges, and potential of IPFS in revolutionizing data storage and management. A significant contribution of this paper is the development of a proof-of-concept web application that employs IPFS for secure and efficient data handling. This application serves as a practical illustration of integrating IPFS into real-world data management scenarios. The security and performance of the application within the decentralized IPFS framework are thoroughly assessed. The study highlights the strengths of IPFS in ensuring data integrity and privacy while acknowledging the challenges in scalability and performance, particularly in handling large files and addressing WebRTC-TCP (Web Real-time Communication-TCP) socket incompatibility issues. Furthermore, we present recommendations for future enhancements of the proof-of-concept web application. These include improving direct file transfer capabilities, advancing file handling techniques, integrating robust key management solutions, and developing dynamic data replication strategies. The research in this paper underscores the potential of decentralized systems like IPFS in shaping the future of data storage, offering a more secure, private, and user-centric approach.**

Keywords-IPFS; Data Privacy; Cloud Storage; Decentralized Storage.

I. Introduction

The field of data storage and access is experiencing a rise in popularity of decentralized models. Centralized data management systems, while established and efficient, present limitations in security, privacy, and user autonomy. This work examines the transition towards decentralized systems, with a focus on the IPFS and distributed data management principles. The motivation for this study arises from an increasing need for secure, private, and user-centric data management solutions. The research addresses several questions:

1) How do decentralized systems like IPFS compare with traditional centralized storage solutions in terms of efficiency, security, and data integrity?
2) What are the main challenges associated with the implementation and use of decentralized storage systems?
3) How can web applications effectively integrate decentralized systems like IPFS for data management, and what are the associated challenges and security implications?

The remainder of the paper is organized as follows. In Section II, we provide some required background and relevant related work, in Section III, we provide details on the design and implementation, whereas in Section IV the approach is evaluated with respect to security and performance. Section V provides some ideas for future work.

II. Background & Related Work

A. Decentralized File Systems

Decentralization in computing and in the web represents a shift from centralized to distributed control. This shift is not merely technical but also philosophical, highlighting ideas such as autonomy, resistance to censorship, and enhanced robustness against failures or attackers. In the early days of computing, centralized systems dominated due to their simplicity. However, the inherent drawbacks, such as single points of failure, scalability issues, and potential for abuse of power led to the exploration of decentralized alternatives [1].

The concept of decentralization in computing started taking shape with early developments. A pivotal development in this direction was the emergence of Peer-to-Peer (P2P) networks, characterized by their lack of reliance on central servers. Napster, one of the first widely used P2P networks, facilitated file sharing by allowing direct file transfers between users' computers. Despite its legal controversies, Napster demonstrated the potential for efficient, decentralized data distribution. Similarly, BitTorrent further advanced this model, efficiently handling large files and numerous simultaneous uploads and downloads, a significant step towards practical decentralized data sharing [2].

The advent of blockchain technology represented a critical development in decentralized systems. Blockchain's introduction of a popular tamper-proof ledger without central authority was first successfully implemented by Bitcoin, the initial decentralized digital currency. This implementation of blockchain technology demonstrated the feasibility of achieving consensus in a trustless environment Subsequently, the development of platforms such as Ethereum expanded the blockchain's applicability. Ethereum introduced functionalities like smart contracts, which allowed for a broader range of decentralized applications, illustrating the versatility and potential of blockchain technology in various domains [3].

B. The InterPlanetary File System

IPFS is a protocol designed to create a peer-to-peer method of storing and sharing media in a distributed file system. Developed as a response to the limitations of the traditional centralized web storage model, IPFS represents a paradigm shift in how information is distributed and accessed [4].

A defining aspect of IPFS is its decentralized nature. Unlike conventional web storage solutions, which rely on centralized servers, IPFS distributes data across a network of nodes. This distribution of data not only mitigates risks associated with single points of failure but also enhances data accessibility and data permanence. Central to IPFS's functionality is content addressing. Traditional web uses location-based addressing, for example, URLs pointing to specific server addresses. In contrast, IPFS addresses content through its content itself by utilizing cryptographic hashing. This approach results in unique content identifiers (CIDs), making content retrieval more efficient and less redundant. Compared to location-based addressing, this method significantly improves both the efficiency and security of data storage and access. By relying on the content's cryptographic hash, immutability is inherent, allowing for verifiable data integrity and significantly contributing to the system's overall robustness [4].

Security in decentralized systems presents a unique set of challenges and considerations distinct from those in traditional centralized architectures. The decentralized nature, while offering advantages in terms of redundancy and resistance to certain types of attacks, also introduces specific vulnerabilities that must be addressed [5], [6].

a) General Security Challenges in Decentralized Systems: Decentralized systems face distinct security vulnerabilities. One key issue is the increased attack surface due to the distributed nature of these systems. Each node in a decentralized network can potentially become a target for attacks. Furthermore, in public decentralized systems, such as IPFS each participating node can potentially be malicious. To ensure that data remains unaltered and private over a distributed network robust encryption and validation mechanisms are required. However, implementing these effectively in a decentralized context, where control is inherently distributed, presents unique challenges in itself [7].

b) Vulnerabilities in DHT-Based Routing Protocols: Distributed Hash Table (DHT) based routing protocols, such as IPFS, have their own vulnerabilities. These include Sybil attacks, where an attacker tries to create a large number of fake identities/nodes to gain a disproportionately large influence on the network [8], [9], and Eclipse attacks, where the attacker isolates a single node or user from the rest of the network, potentially feeding it false and/or harmful information.

c) Network Reliability: Maintaining a consistent and reliable network is another critical challenge in decentralized systems. In IPFS, for instance, the absence or unavailability of nodes can lead to difficulties in data retrieval, highlighting the need for robust network health.

d) Data Persistence and Redundancy: In decentralized systems like IPFS, data persistence is dependent on nodes electing to store that data. Unlike centralized systems where data storage can be systematically managed and guaranteed, IPFS faces the challenge of ensuring that data remains available even when the node originally providing that data goes offline. This issue necessitates a redundant storage mechanism and an incentive for nodes to retain data.

III. Approach

The web application developed as part of this work represents a proof-of-concept for a secure, decentralized file storage system. It operates within the broader ecosystem of IPFS, leveraging the decentralized nature of the platform to offer a novel approach to data storage and access. The application is hosted directly on IPFS, which provides a resilient and distributed hosting solution. This hosting choice aligns with the overarching theme of decentralization, ensuring that the application itself is as robust and distributed as the data it manages.

A. Attacker models

In this work, we focus on the following three attacker models, as we consider them to be the most important ones with respect to IPFS:

1) Malicious IPFS Nodes: Given the open nature of IPFS, the application may interact with nodes that attempt to access or manipulate user data. To mitigate this, the application employs end-to-end encryption, ensuring that data remains secure and unreadable by unauthorized nodes.
2) Data Manipulation Attacks: The possibility of an attacker altering the data in transit is addressed through the use of IPFS's content addressing and the application's encryption mechanisms. The integrity of data is maintained as any alteration in the encrypted data will be detectable due to the change in its CID. Contrary to the next attack, this attacker might only try to redirect traffic or alter information, even without being able to actually decode it.
3) Eavesdropping Attacks: The risk of data interception is countered by encrypting the data before it is shared or stored on the network. This ensures that even if the data is intercepted, it remains incomprehensible to the attacker. Contrary to the previous attacker, this attacker is only passively involved, i.e. he/she does not change data.

The logic behind the selection of these three models is that one is an attack from inside the network, namely the most prominent one where a node is made malicious, while attacker model two and three model an active, as well as a passive, attacker respectively.

B. Connectivity and File Processing Framework

In the current IPFS ecosystem, direct file transfers using the Helia library face challenges due to a WebRTC-TCP incompatibility issue in current IPFS nodes. As a pragmatic approach, the application utilizes third-party HTTP APIs for interactions with storage providers like Filebase or Pinata. This strategy is a temporary solution until direct transfer of files via IPFS becomes feasible. The use of HTTP APIs as a current means of file handling offers reliable storage and retrieval, albeit with a modest departure from the ideal decentralized model [10].

C. Custom Identity Management and File Synchronization

This section elaborates on the unique approach to identity management and file synchronization within the developed web application. The system hinges on the creation and utilization of a user-specific identity file, coupled with a dynamic file indexing mechanism, ensuring secure and efficient interactions with the IPFS network.

a) Identity File Creation and Usage: The first step is the creation of an identity file for the user. This pivotal process involves:

- Generation of an AES (Advanced Encryption Standard) private key, either supplied by the user or automatically generated by the application.
- Requirement for the user to input an API key from a chosen third-party storage provider. This is required due to current Helia limitations (WebRTC - TCP incompatibility).

The identity file, essentially a JSON object, encompasses crucial components for user identification and interaction with the IPFS network:

- The IPNS (InterPlanetary Name System) name, pointing to the file index JSON object
- The user's AES private key
- The user's third-party API key

This identity file represents the core of user data portability, enabling access to their IPFS storage from any device by merely transferring this file.

b) File Structure and Index File Mechanism: Every file, including the index file, adheres to a structured format:

- Composed as JSON objects
- Contains encrypted data as a Uint8-Array
- Includes the AES-GCM initialization vector it was encrypted with

Upon uploading the first file, the index file is generated, and an IPNS entry is created to consistently point to the latest CID of this index file. The index file plays a crucial role in the system:

- Structured as a JSON object.
- Contains an encrypted list of all files, each entry detailing:
  - File CID
  - File name
  - File size
  - SHA-256 hash of the file
  - Optional metadata for enhanced file information (e.g. a timestamp of the last change).

Currently, this approach does not account for collisions, as 256 bit hashes have a wide result space, thus the probability of accidental collisions is very low. Still, this could be improved in future versions.

c) File Retrieval Process: The steps to retrieve a file are as follows:

1) Connect to IPFS and query the CID of the index file JSON object.
2) Download the index file JSON object.
3) Decrypt the file list using the attached AES-GCM initialization vector and the user's AES256 private key.
4) Display metadata of all files contained in the index.
5) On file request, query the specific CID.
6) Download the requested file JSON object.
7) Decrypt the file using the attached AES-GCM initialization vector and the user's AES256 private key.

d) File Storage Process: When a new file is uploaded or an existing one modified, the following steps are undertaken:

1) Encrypt the file using the user's AES256 private key and a newly generated initialization vector.
2) Create a JSON object for the file, storing the encrypted data and the initialization vector.
3) Upload the file to an IPFS storage provider (using Helia or third-party HTTP APIs).
4) Record the file's CID and metadata, appending it to the index file.
5) Request the storage provider to pin the new file on IPFS.
6) Upon successful pinning, upload the updated index file and request pinning.
7) Update the IPNS entry to reflect the new index file.
8) Unpin the old index file (and the old file if it was an update) after successful IPNS update and new file pinning.

This architecture ensures a secure, user-friendly, and efficient mechanism for managing files on the decentralized IPFS network, addressing current limitations while laying the groundwork for future improvements in direct file transfer capabilities.

D. Encryption in the Application

The approach implements AES-GCM, an Advanced Encryption Standard in Galois/Counter Mode, for its data encryption and decryption processes. AES-GCM is chosen primarily due to its integration with the Web Crypto API, along with its Authenticated Encryption with Associated Data (AEAD) properties and widespread hardware acceleration support. The selection of AES-GCM for encryption in the application is driven by several factors:

- Web Crypto API Compatibility: AES-GCM is readily available in theWeb Crypto API, facilitating easy implementation in web applications [11].
- AEAD: AES-GCM provides both encryption and data integrity, ensuring data confidentiality and protection against tampering [12].
- Hardware Acceleration: The widespread hardware support for AES that allows for fast computation with cheap hardware.

IV. Evaluation and Conclusions

A. Security evaluation

The web application was designed with specific attacker models in mind, primarily focusing on safeguarding user data from unauthorized access and manipulation. With respect to the attacker models outlined in Section III, the following conclusions could be drawn:

1) Malicious IPFS Nodes: The primary threat comes from malicious nodes within the IPFS network that may attempt to access or tamper with user data. The application's use of AES-GCM encryption effectively counters this threat by ensuring data confidentiality. Encrypted files, even if intercepted, remain inaccessible to unauthorized parties.

2) Data Manipulation Attacks: Another concern is the potential for data manipulation during transmission. The self-verifying nature of IPFS CIDs, combined with the integrity assurance of AES-GCM, provides robust protection against such attacks. This dual layer of security ensures that any tampered data is easily detectable.
3) Sybil and Eclipse Attacks: While the application does not directly mitigate DHT vulnerabilities like Sybil and Eclipse attacks, it minimizes their impact on user data privacy. The encrypted data stored on IPFS remains secure against these attacks, as the encryption layer acts independently of the underlying DHT's vulnerabilities [7], [9].

In addition, the use of AES-GCM for encryption plays a crucial role in securing user data:

- Data Confidentiality: AES-GCM ensures that file contents remain confidential. By encrypting data before it is uploaded to IPFS, the application prevents unauthorized access, even if the data is replicated across potentially untrustworthy nodes.
- Data Integrity: Alongside confidentiality, AES-GCM provides data integrity checks. This feature is critical in a decentralized setting where data passes through multiple nodes, as it enables the detection and rejection of tampered data.
- Performance Considerations: While AES-GCM is computationally efficient due to widespread hardware acceleration support, the application's encryption process is dependent on the user's device capabilities. This can impact performance, particularly for larger files.

In conclusion, the security evaluation reveals that the web application effectively addresses key security concerns within the decentralized IPFS framework. The robust encryption strategy ensures data confidentiality and integrity, mitigating risks associated with decentralized data storage and transmission. The application's current security measures provide a solid foundation, though future enhancements could focus on advanced key management and addressing broader DHT vulnerabilities.

B. Performance and Stability

This section focuses particularly on efficiency in handling large files, WebRTC and TCP socket compatibility issues, and the implications of these factors on data loss prevention and redundancy strategies.

1) Efficiency and Large File Handling: The application's current architecdture faces challenges in managing large files due to limitations in the splitting and handling of large data sets. Key observations include that large files lead to extended encryption and upload times, constrained by the device's RAM and processing power. Furthermore, there are also some concerns regarding scalability: Without the ability to split large files into manageable blocks, the application's scalability is hindered, particularly when dealing with extensive data sets or high-volume storage requirements.

2) WebRTC and TCP Socket Incompatibility: One of the primary limitations in the current implementation of the application is the incompatibility between WebRTC and TCP sockets within the IPFS ecosystem. This limitation impacts the stability of the application. Regarding connection limitations, due to this incompatibility, the application primarily relies on a few nodes that act as gateways for browser-based interactions. This reliance can lead to bottlenecks and potential points of failure [10]. Furthermore, the reliance on HTTP APIs could be a problem: The application currently uses HTTP APIs of third-party IPFS storage providers like Filebase or Pinata for file handling, which, while reliable, deviates from the ideal decentralized model and could impact long-term scalability and decentralization goals [10].

3) Data Loss Prevention and Redundancy: In addressing the concerns of data loss and ensuring redundancy, the application leverages the inherent strengths of the IPFS network. Regarding the decentralized storage, data availability in IPFS is independent of the storage location, allowing for convenient replication and enhanced scalability. Utilizing crypto-based storage providers like Filecoin offers a cost effective solution for redundant storage. At the time of writing, the cost of storing 10TB of data on Filecoin (1.95 USD per month) is significantly lower than traditional cloud storage options like Amazon S3 (235 USD per month) [13], [14]. Regarding collateral-based reliability, Filecoin storage providers are required to provide collateral, adding an additional layer of reliability and commitment to data preservation [15].

In conclusion, the evaluation of the application's performance and stability highlights key areas for improvement, particularly in large file handling and direct file transfer capabilities. Despite these challenges, the application benefits from the decentralized, scalable nature of IPFS and the cost-effective, redundant storage solutions offered by crypto-based storage providers. Future work should focus on enhancing file transfer capabilities and exploring more efficient file processing methods to bolster the application's performance and scalability within the decentralized web.

V. Future Work

Based on the findings, several recommendations are proposed to advance the field in terms of future work: Regarding technical improvements, a lot of improvements could be done by addressing the limitations identified in IPFS, particularly in file transfer and with respect to compatibility. Regarding key management, fruitful future work lies in investigating more sophisticated encryption key management techniques, including hardware solutions like TPM2 [16], which could significantly improve security. Regarding automated data replication, developing mechanisms for user-defined data replication will increase redundancy and reliability, especially using cost-effective storage solutions like Filecoin.

References

[1] M.-Å. Hugoson, "Centralized versus decentralized information systems: A historical flashback", in *History of Nordic Computing 2: Second IFIP WG 9.7 Conference, HiNC2, Turku, Finland, August 21-23, 2007, Revised Selected Papers 2*, Springer, 2009, pp. 106–115.

[2] P. Raj *et al.*, "High-performance peer-to-peer systems", *High-Performance Big-Data Analytics: Computing Systems and Approaches*, pp. 317–337, 2015.

[3] W. Wang *et al.*, "A survey on consensus mechanisms and mining strategy management in blockchain networks", *Ieee Access*, vol. 7, pp. 22 328–22 370, 2019.

[4] IPFS, "Official ipfs documentation", retrieved: March 2025, [Online]. Available: https://docs.ipfs.tech/.

[5] E. Karaarslan and E. Konacaklı, "Data storage in the decentralized world: Blockchain and derivatives", *arXiv preprint arXiv:2012.10253*, 2020.

[6] N. Z. Aitzhan and D. Svetinovic, "Security and privacy in decentralized energy trading through multi-signatures, blockchain and anonymous messaging streams", *IEEE transactions on dependable and secure computing*, vol. 15, no. 5, pp. 840–852, 2016.

[7] B. Prünster, A. Marsalek, and T. Zefferer, "Total eclipse of the heart–disrupting the {interplanetary} file system", in *31st USENIX Security Symposium (USENIX Security 22)*, 2022, pp. 3735–3752.

[8] L. Wang and J. Kangasharju, "Real-world sybil attacks in bittorrent mainline dht", in *2012 IEEE Global Communications Conference (GLOBECOM)*, IEEE, 2012, pp. 826–832.

[9] J. R. Douceur, "The sybil attack", in *International workshop on peer-to-peer systems*, Springer, 2002, pp. 251–260.

[10] ipfs-helia, "Helia github repository.", retrieved: March 2025, [Online]. Available: https://github.com/ipfs/helia/issues/256.

[11] Mozilla Foundation, "Web crypto api", retrieved: March 2025, [Online]. Available: https://developer.mozilla.org/en-US/docs/Web/API/Web_Crypto_API.

[12] D. McGrew, "An interface and algorithms for authenticated encryption", Tech. Rep., 2008.

[13] Amazon Web Services, "Amazon s3 pricing", retrieved: April 2024, [Online]. Available: https://aws.amazon.com/s3/pricing/.

[14] Storage.market, "Ipfs storage market", retrieved: April 2024, [Online]. Available: https://file.app/.

[15] Filecoin, "Official filecoin documentation", retrieved: March 2025, [Online]. Available: https://docs.filecoin.io/.

[16] K. Shang *et al.*, "Cluster nodes integrity attestation and monitoring scheme for confidential computing platform", in *2023 IEEE 22nd International Conference on Trust, Security and Privacy in Computing and Communications (TrustCom)*, IEEE, 2023, pp. 740–749.